Current Topics in Plant Physiology:
An American Society of Plant Physiologists Series
Volume 4

Calcium in Plant Growth and Development

Edited by
Robert T. Leonard
Peter K. Hepler

Proceedings
13th Annual Riverside
Symposium in Plant Physiology
January 11-13, 1990

Department of Botany and Plant Sciences
University of California, Riverside

Published by:
American Society of Plant Physiologists
15501-A Monona Drive
Rockville, Maryland 20855

Copyright 1990. All rights reserved. No part of this publication may be reproduced without the prior written permission of the publisher.

Library of Congress Cataloging in Publication Data

Main entry under title:

Calcium in Plant Growth and Development

Current Topics in Plant Physiology: An American Society of Plant Physiologists Series, Volume 4.
Includes bibliographies and index.

1. Calcium--Congresses. 2. Plant-Growth--Congresses. 3. Plant Development--Congresses.

I. Leonard, Robert T., 1943- . II. Hepler, Peter K., 1936- III. University of California, Riverside. Dept. of Botany and Plant Sciences. V. Title.

Library of Congress Catalog Card Number: 89-082676
ISBN 0943088-18-6

Standing (Left to Right)
C. Lagarias, M. Harrington, A. Bennett, R. Leonard, R. Cleland, A. Läuchli, I. Ferguson, F. DuPont, M. Evans

Sitting (Left to Right)
D. Briskin, R. Williamson, R. Wayne, S. Wick, P. Hepler, L. Jaffe, F. Harold, L. Taiz, K. Robinson

Standing (Left to Right)

R. Bruce, C. Colling, R. Zielinski, A. Roberts, L. Klimczak, D. Brauer, R. Borchert, M. Webb, I. Ferguson, C. Haigler, D. Kropf, A. Wrona, M. Jaffe, S. Bartnicki-Garcia

Sitting (Left to Right)

F. DuPont, J. Garbarino, R. Chaubal, M. Lazzaro, K. Ketchum, M. Chou, C. Reiss, S. Snapp, H. Ishikawa

CONTRIBUTORS

J. R. Aist
Melanie J. Barnett
S. Bartnicki-Garcia
Swati Basu
A. B. Bennett
Tom Berkelman
P. Bernasconi
Donald Bers
Thomas Björkman
Rolf Borchert
David Brauer
Donald P. Briskin
Robert J. Bruce
Maria Burchert
Douglas S. Bush
Dale A. Callaham
Rajendra Chaubal
Robert E. Cleland
Christiane Colling
Mark Collinge
Frances M. DuPont
David W. Ehrhardt
Michael L. Evans
N. N. Ewing
I. B. Ferguson
J. E. Garbarino
G. Gierz
Lynne H. Gildensoph
J. P. Gogarten
R. Goren
Franz Grolig
Ruth P. Hagan
Susumu Hagiwara
Klaus Hahlbrock
Candace H. Haigler
H. Michael Harrington
Peter K. Hepler
F. Hergert

G. Hind
Susan R. Hurst
Hideo Ishikawa
Peter P. Jablonsky
Lionel F. Jaffe
M. J. Jaffe
Russell L. Jones
Heinrich Kauss
Karen A. Ketchum
H. Kibak
Helen G. Kiss
L. J. Klimczak
Georg Kreimer
Darryl L. Kropf
J. Clark Lagarias
Erwin Latzko
André Läuchli
Mark D. Lazzaro
Vincent Ling
Sharon R. Long
Ying Tang Lu
Stefan Moisyadi
Yuli Moriyasu
Imara Perera
Ronald J. Poole
T. Rausch
Bonnie J. Reger
Carol Reiss
Kenneth R. Robinson
Alison W. Roberts
Dierk Scheel
Julian I. Schroeder
C. Schubert
C. Shennan
J. P. Slattery
S. S. Snapp
R. M. Spanswick
Mark Staves

I. Struve
Barbara Surek
L. Taiz
S. L. Taiz
Douglas Taylor
William W. Thomson
S-I Tu
B. Veierskov
Sarbjit S. Virk
D. Martin Watterson
Randy Wayne
Mary Alice Webb
Charles A. West
Susan Wick
Richard E. Williamson
Emily Wilson
L. E. Wimmers
A. F. Wrona
Dahong Zhang
Raymond E. Zielinski

EDITORS' COMMENTS

INTRODUCTION

In the last decade, there has been a markedly heightened interest in calcium with the realization that the ion may form a crucial link between stimulus and response in a variety of different physiological and developmental processes. Studies on animal systems have made enormous strides in deciphering the details and the complexities of the calcium messenger pathways, and have provided the paradigms that guide current work. By comparison, progress on these problems in plant systems has been slow, and relatively few signal pathways have been elucidated. However, plants are different in many fundamental ways from animals and, not surprisingly, their processes of calcium utilization and signal transduction may also be different. Nevertheless, the likelihood that the ion occupies a pivotal role in several different signal transduction processes, as in animal systems, appears assured. It seemed timely, therefore, to convene a meeting that focuses on the calcium question as it applies to plants, both as a way to assess what has been learned, and to provide direction for future research.

In developing the program for the meeting, we attempted to broadly cover the subject area from the role of calcium as a nutrient element, to its activity as a second messenger. Some of the questions that seemed important to address included the following: What is the significance of calcium in the cell wall? What are the pathways of calcium transport into the cell down its free energy gradient, and out of the cytoplasm against its free energy gradient? What is the activity of the free ion in the cytoplasm? Does this activity change in response to a signal? If it does change, what and where are the target proteins and reactions that are stimulated or inhibited? Ultimately, what processes are regulated? Finally, do local gradients in the free calcium concentration exist, and what are their roles in plant growth and development?

It is exciting to see that good progress is beginning to emerge on some of these questions. It is perhaps even more encouraging to realize that the important groundwork, which is essential for major advances in the future, is being established. For example, many of the crucial experimental approaches that have proved extremely powerlful in the solution of signal transduction problems in animal systems are being adapted for work on plant systems. Thus, the modern techniques of molecular biology, electrophysiology, and fluorescence ratio imaging are being used along side of the more conventional methods of biochemistry, physiology and microscopy to provide answers to difficult but important questions. There is reason, therefore, to be optimistic that the next decade will provide a greatly expanded understanding of the contribution of calcium to a myriad of processes that underlie fundamental aspects of plant growth and development.

DEDICATION TO RUTH SATTER

In developing a list of participants for this symposium, we had very much wanted to include Ruth Satter. She and co-workers had recently made excellent progress in deciphering key aspects of the signal transduction pathway involved in the control of circadian leaf movements in *Samanea saman*. Specifically, they found that brief illumination with light stimulated the breakdown of membrane-associated phosphatidyl bisphosphate, and generated an increase in inositol trisphosphate. Since the latter is presumed to cause calcium release from internal stores, her work came to occupy a position in the forefront of our thinking about how plant cells modulate intracellular calcium ion activity.

Although we knew that Ruth was ill, she never told anyone about the nature of her illness. We asked her to participate as a speaker in the conference, but she declined saying that she did not think it was wise for her to travel. Sadly, in early August 1989, she died. It then became known to us that she had been fighting leukemia for nine years. In retrospect, one must look back and stand in awe of the truly remarkable strides that Ruth made during those nine years. Of course, she was already recognized as a highly respected, productive plant physiologist, clearly one of the leaders of circadian rhythm research. But during these last years, while maintaining her already active program, she ventured into new areas that included not only the polyphosphoinositide biochemistry mentioned above, but patch-clamp electrophysiology. In addition, between January and July, 1989, she and co-workers conceived and brought to fruition, a book entitled "The Pulvinus: Motor Organ For Leaf Movement."

For those of us who have known Ruth, her special qualities included a rare combination of originality, intelligence and determination, together with personal warmth and generosity. It is with the greatest admiration and respect that we dedicate this volume to the memory of Ruth Satter.

ACKNOWLEDGMENTS

The editors want to acknowledge Cindi McKernan of the Department of Botany and Plant Sciences at the University of California, Riverside (UCR) for her many and varied contributions to the organization of the Symposium. She so expertly handles the many details required to bring scientists together and she does so without intruding on the scientific goals of the meeting. We also appreciate the expert advice on editing provided by Patti Fagan and Aileen Wietstruk, and their skillful preparation of the camera-ready copy of this volume. We would also like to acknowledge the graduate students, postdoctorals, and staff members of the Department of Botany and Plant Sciences for their assistance with various aspects of the Symposium.

We respectfully thank the following sponsors for their financial support: The Cooperative State Research Service of the United States Department of Agiculture; the Department of Botany and Plant Sciences, UCR; Chancellor Rosemary S. J. Schraer, UCR; Seymour D. Van Gundy, Dean of the College of Natural and Agricultural Science, UCR; and The University of California Biotechnology Research and Education Program. We are also grateful for financial support from the American Society of Plant Physiologists, Beckman Instruments, Inc., Monsanto Co., and Sandoz Crop Protection Corporation.

Finally, we sincerely thank the distinguished scientists who traveled from near and far to attend and present papers during this Symposium.

Peter K. Hepler
Robert T. Leonard

FOREWORD

With this volume, the American Society of Plant Physiologists continues its series of publications on timely topics in plant physiology. Publication of proceedings devoted to focus areas, such as the present one on calcium in plant growth and development, is designed to share information from the symposia with other scientists. This book is the fourth in the series "Current Topics in Plant Physiology: An American Society of Plant Physiologists Series." It is the wish of the Publications Committee and the Executive Committee of the Society to make these publications as useful as possible. To this end, copies of this publication and publications from previous years are available at an affordable price from the American Society of Plant Physiologists, 15501 Monona Drive, Rockville, Maryland 20855, telephone 301/251-0560.

The ASPP Publications Committee
Machi F. Dilworth, Chair
Gerald E. Edwards   Chris R. Somerville
James E. Harper   Joseph E. Varner

Previous titles in the series are:

Volume 3, 1990: THE PULVINUS: MOTOR ORGAN FOR LEAF MOVEMENT,
Eds R. L. Satter, H. L. Gorton, T. C. Vogelmann

Volume 2, 1989: PHYSIOLOGY, BIOCHEMISTRY, AND GENETICS OF NONGREEN PLASTIDS, Eds C. D. Boyer, J. C. Shannon, R. C. Hardison

Volume 1, 1989: PLANT REPRODUCTION: FROM FLORAL INDUCTION TO POLLINATION,
Eds E. M. Lord, G. Bernier

Included among earlier ASPP symposium publications are:

1988: LIGHT-ENERGY TRANSDUCTION IN PHOTOSYNTHESIS: HIGHER PLANT AND BACTERIAL MODELS, Eds S. E. Stevens, Jr., D. A. Bryant

1988: PHYSIOLOGY AND BIOCHEMISTRY OF PLANT MICROBIAL INTERACTIONS, Eds N. T. Keen, T. Kosuge, L. L. Walling

1987: PLANT SENESCENCE: ITS BIOCHEMISTRY AND PHYSIOLOGY,
Eds W. M. Thomson, E. A. Nothnagel, R. C. Huffaker

1987: PHYSIOLOGY OF CELL EXPANSION DURING PLANT GROWTH,
Eds D. J. Cosgrove, D. P. Knievel

1986: REGULATION OF CARBON AND NITROGEN REDUCTION AND UTILIZATION IN MAIZE, Eds J. C. Shannon, D. P. Knievel, C. D. Boyer

1986: MOLECULAR BIOLOGY OF SEED STORAGE PROTEINS AND LECTINS,
Eds L. M. Shannon, M. J. Chrispeels

1986: EXPLOITATION OF PHYSIOLOGICAL AND GENETIC VARIABILITY TO ENHANCE CROP PRODUCTIVITY, Eds J. E. Harper, L. E. Schrader, R. W. Howell

1985: INORGANIC CARBON UPTAKE BY AQUATIC PHOTOSYNTHETIC TISSUE,
Eds W. J. Lucas, J. A. Berry

CONTENTS

Contributors vii

Editor's Introduction viii

Foreword x

Contents xi

List of Abbreviations xiv

Session I - Calcium as a Mineral Nutrient

- Calcium Nutrition and Cellular Response 1
    I. B. Ferguson

- Calcium, Cell Walls and Growth 9
    Robert E. Cleland, Sarbjit S. Virk,
    Douglas Taylor, and Thomas Björkman

- Calcium Binding by Chloroplast Stroma Proteins
    and Functional Implications 17
    Maria Burchert, Barbara Surek, Georg Kreimer,
    and Erwin Latzko

- Calcium, Salinity and the Plasma Membrane 26
    André Läuchli

Session II - Compartmentation and Transport of Calcium in Plant Cells

- Molecular Biology of Plant P-Type Ion-Translocating ATPases 36
    L. E. Wimmers, N. N. Ewing, and A. B. Bennett

- Characterization of the Ca^{2+}-Transporting ATPase of the
    Plant Plasma Membrane Using Isolated
    Membrane Vesicles 46
    Donald P. Briskin, Lynne H. Gildensoph, and Swati Basu

- The Vacuolar ATPase: Structure, Evolution, and
    Promoter Analysis 55
    L. Taiz, I. Struve, T. Rausch, P. Bernasconi,
    J. P. Gogarten, H. Kibak, and S. L. Taiz

- Hormonal Regulation of Ca^{2+} Transport in Microsomal
   Vesicles Isolated from Barley Aleurome Layers 60
   Douglas Bush and Russell L. Jones

Session III - Calcium and Cellular Movements

- The Role of Calcium in Muscle Physiology 67
   Donald Bers

- Immunolocalization of Chara Calmodulin and the Reversibility of
   the Inhibition of Cytoplasmic Streaming by Ca^{2+} 79
   Peter P. Jablonsky, Ruth P. Hagan,
   Franz Grolig, and Richard E. Williamson

- Calcium, Cytoplasmic Streaming, and Gravity 86
   Randy Wayne, Mark Staves, and Yuli Moriyasu

- Calcium and the Regulation of Mitosis 93
   Peter K. Hepler, Dahong Zhang, and Dale A. Callaham

Session IV - Calcium as a Second Messenger

- Temporal and Spatial Changes in Ca^{2+} during
   Plant Development 111
   Kenneth R. Robinson

- Calcium Ion Currents and Gradients in Fucoid Eggs 120
   Lionel F. Jaffe

- A Molecular Genetics and Mutant Analysis Approach for
   Elucidation of Molecular Mechanisms of
   Calcium Signal Transduction through
   Calmodulin:Calmodulin-Binding Protein Complexes 127
   Emily Wilson, D. Martin Watterson, and Mark Collinge

- Localization of Calcium-Binding Proteins in Dividing Plant Cells 137
   Susan Wick

Session V - Calcium and Responses to the Environment

- Voltage-Dependent Activation of Ca^{2+}-Regulated Anion Channels
   and K^+ Uptake Channels in *Vicia Faba* Guard Cells 144
   Julian I. Schroeder and Susumu Hagiwara

- Calcium Transport in the Unicellular Green Alga *Mesotaenium Caldariorum*: A Model System for Phytochrome-Mediated Responses 151
    Tom Berkelman and J. Clark Lagarias

- Structural and Functional Analysis of Heat Shock Proteins 161
    H. Michael Harrington, Stefan Moisyadi, and Ying Tang Lu

- Interaction of Calcium and Auxin in the Regulation of Root Elongation 168
    Michael L. Evans, Helen G. Kiss, and Hideo Ishikawa

Poster Abstracts 176

Index 201

ABBREVIATIONS

A-9-C	9-Anthracenecarboxylic acid
AM	Acetoxy Methylester
APK	autophosphorylating protein kinase
APW	artificial pond water
BAPTA	1,2-bis(*O*-aminophenoxy)ethane-*N,N,N',N'*-tetraacetic acid
BTP	bis-tris propane
Ca	Calcium
Ca_i^{2+}	intracellular-free calcium
CaM	calmodulin
CaMBP	calmodulin-binding protein
CAT	chloramphenicol acetyltransferase
CHAPS	(3-[cholamidopropyl)-dimethylammonio]-1-propanesulfonate
CTC	chlorotetracycline
DAG	diacylglycerol
DCCD	N,N'-dicyclohexylcarbodiimide
DES	diethylstilbestrol
DFS	Donnan Free Space
DHP	dihydropyridine
E-C	excitation contraction
E-C coupling	effect of extracellular Ca^{2+} on streaming in response to an action potential
EITC	erythrosin isothiocyanate
E_m	membrane potential
FBPase	fructose-1,6-bisphosphatase [EC 3.1.3.1]
FS	Free Space
Fura-2	{1-[2-(5-carboxyoxazol-2yl)-6-aminobenzofuran-5-oxy]-2-(2-amino-5'-methylphenoxy)-ethane-N,N,N',N'-tetraacetic acid}
GA	gibberellic acid
H_{II}	hexagonal II
HS	heat shock
HSP	heat-shock protein
HSR	heat-shock response
InsP3	inositol triphosphate
IP_3	inositol-(1,4,5)-trisphosphate
K_D	dissociation constants
LMWC	low mol wt complex
MPF	mitosis-promoting factor
MT	microtubule
MTOC	microtubule organizing center

nt	normal Tyrode's solution
PEx	instron plastic
PI	phosphoinositol
PIP	phosphoinositol phosphate
PIP_2	phosphatidylinositol bisphosphate
PtdIns	phosphatidylinositol
PtdInsP	phosphatidylinositol monophosphate
PtdInsP2	phosphatidylinositol bisphosphate
Quin-2	2-((2-bis-(carboxymethyl)amino-5-methylphenoxy)methyl)-6-methoxy-8-bis(carboxymethyl)aminoquinoline
RCC	rapid-cooling contractures
R/FR	red/far red
SBPase	sedoheptulose-1,7-bisphosphatase [EC 3.1.3.37]
$S_{K,Na}$	selectivity coefficient
SR	sarcoplasmic reticulum
TEA	tetraethylammonium chloride
V-ATPase	vacuolar proton-pumping ATPase

CALCIUM NUTRITION AND CELLULAR RESPONSE

I. B. Ferguson

*DSIR Fruit & Trees, Mt Albert Research Centre
Private Bag, Auckland, New Zealand*

Calcium nutrition of plants has always provided special features which have beguiled plant physiologists over the decades. If deficiencies are studied, then it is clear that those associated with Ca are complex and often indirect. If Ca is studied in terms of mechanisms, then it stands out amongst the major nutrients for its extracellular transport pathways, and problematic symplastic movement. A consideration of its physiological role highlights a need for reconciliation of recent findings on the cellular regulation of Ca and its function in cell response with our more established views on Ca as a nutrient.

The main aspects of Ca as a nutrient, which have governed our study of the mineral in the past, are: (*i*) soil availability, (*ii*) root uptake, (*iii*) transport and translocation, (*iv*) tissue content and distribution, (*v*) nutrient interaction, (*vi*) cell content and distribution, and (*vii*) deficiencies. These aspects, in fact, have told us less about the role of Ca in plant tissues than about the characteristics of Ca uptake and transport in the plant. The two important conclusions reached from the nutritional standpoint are that Ca is distributed primarily with the water flow along apoplastic pathways, and that the carefully regulated cellular distribution of Ca is the key to its physiological function.

The traditional functions of Ca have revolved around involvement in cell wall structure, and in membrane structure and function. The most recent comprehensive reviews on Ca as a nutrient in plants (*e.g.* 8) have however, foreseen the secondary messenger functions. While the broad conclusions on Ca nutrition made over the last decade or so are still applicable, they are worth considering further in reconciling our older views of Ca nutrition with more recent information on cellular response. Consider some of these aspects further, using the apple fruit as an example. The developing and ripening apple fruit provides some of the best examples of how problems develop with Ca nutrition, not only in relation to the development of a specific Ca deficiency such as bitter pit (5), but also with respect to transport and distribution of Ca in the plant.

TRANSPORT IN THE PLANT

The total Ca content of apple fruit often reaches a maximum level relatively early in fruit development (Fig. 1), whereas Mg and K both continue to move into the fruit over the whole period of growth. Thus, the Ca concentration in the fruit flesh declines over almost the whole period of fruit expansion, and Mg and K concentrations may be maintained (K) or decline to a lesser extent (Mg). The conclusion is that Ca movement into the fruit does not keep up with fruit expansion, and the largely untested reasoning has been that when phloem transport becomes predominant, Ca input into the fruit declines (6). These effects of the transport systems on mineral content of the fruit can also be seen when fruit of different sizes are analysed. Calcium concentration declines with increasing fruit size, whereas concentrations of Mg and K frequently are maintained, or even increase (Table I).

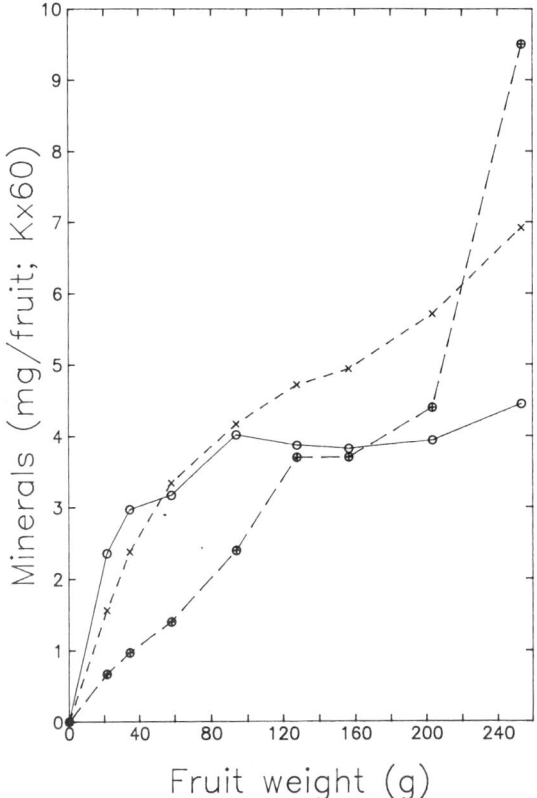

Figure 1. Ca (•), Mg (o) and K (x) contents of Granny Smith apple fruit during fruit growth to maturity and harvest.

Table I. *Mineral concentrations of flesh of mature Cox's Orange Pippin apple fruit.* Data are mean values from 4 orchards, 4 analytical samples being taken per fruit weight class.

Fruit weight	Calcium	Magnesium	Potassium
g		mg/100 g fresh weight	
127	1.69	3.87	138.5
138	1.68	3.92	142.3
152	1.54	4.07	146.1
168	1.52	3.94	149.1

Practically, this means that to ensure the maximal input of Ca into the fruit, the rate of uptake in the early stages of growth must be increased, or input in the later stages of the season must be continued. These needs have led us to look at the effects of leaves associated with the fruiting spur on mineral transport. Contrary to the widespread belief that leaves compete with fruit for minerals, those on the spur itself, either spur leaves or those on bourse shoots, can have positive effects in attracting Ca into the spur, and so enhance the mineral supply (12; Fig. 2). These results serve to emphasize the importance of the driving force of transpiration for the subsequent location and distribution of Ca in plant tissues. The practical importance is particularly evident in organs which do not have efficient transpirational surfaces -- bulky organs such as fruits, meristematic tissues, etc. In such sites, Ca may not reach minimum "critical" concentrations, and cell function may become impaired under various physiological conditions. The result is often a Ca-related disorder such as bitter pit of apples, cork spot of pears, blossom-end rot of tomatoes and blackheart of celery.

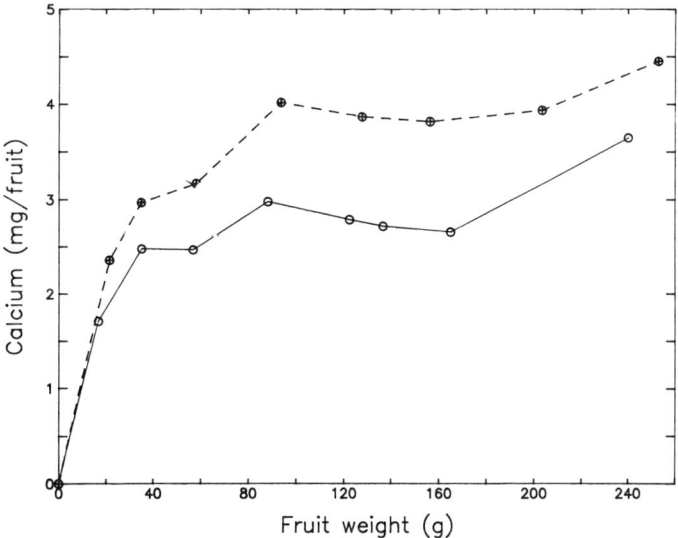

Figure 2. Calcium content of Granny Smith apple fruit during fruit growth on spurs with 21 (•) and 8 (o) leaves per bourse shoot per spur.

There is still often a practical need to know the "critical" concentration of Ca in a particular tissue, but our current knowledge of cellular distribution raises the unresolved question of whether such a concept, however practically useful, has any physiological meaning. Similar doubts are raised by the question of Ca requirements of a plant. There used to be talk of Ca as a micronutrient (19), but physiological advances have largely passed this by.

LOCALIZATION AND DISTRIBUTION

The variable nature of the cellular distribution of Ca -- with differing contents in the vacuole, organelles, cytosol and extracellular fluid -- suggests that whole tissue analysis would not be very useful in providing practical relationships between tissue content and a disorder associated with a Ca deficiency. However, these have been derived in the case of bitter pit in apple fruit (6), and although there is a lot of inherent variability in sampling and analysing for Ca in material such as fruits, a relationship such as shown in Figure 3 is used to predict bitter pit risk in orchards.

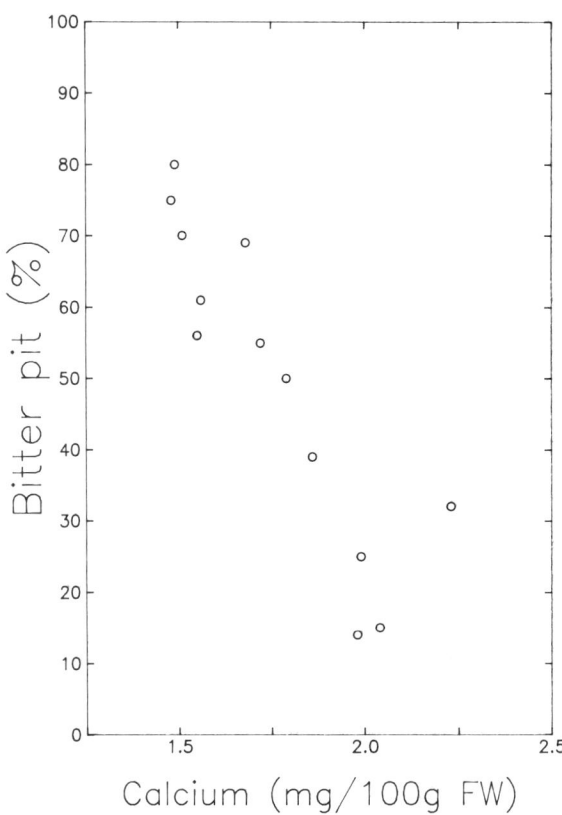

Figure 3. Relationship between bitter pit incidence after storage, and Ca concentration of Cox's Orange Pippin apple fruit. Each point represents fruit from a different orchard.

If we do get a reliable relationship between tissue content and a disorder, does this mean that the major pools of Ca, which are probably the vacuole and the extracellular fluid, are more important or critical than we have thought? Some recent measurements on mature apple fruit using Ca-selective microelectrodes have shown that vacuolar and apoplastic concentrations averaged 0.6 mM and 0.1 mM, respectively, (Harker and Venis, unpublished). Cytosolic concentrations are in the submicromolar range, and thus the vacuole and extracellular fluid provide pools of Ca potentially available for cellular response or other regulatory mechanisms.

The Ca in the apoplast is of particular interest. We have recently shown that Ca movement in the apoplast of apple fruit flesh is by a combination of exchange mechanisms, diffusion, and mass flow (9, 10). At low levels of Ca, cation exchange is the dominant force, and at higher levels, diffusion. As the fruit develops, the relative cell wall surface area and cation exchange capacity of the tissue and cell wall yield decrease, air space increases, and Ca transport rates decrease (9, Fig. 4). If the Ca in the apoplast solution is in any way critical for normal cell function, then changes in the characteristics of the apoplast, either those affecting transport, or those such as the Ca binding and buffering capacity of the cell wall which affect Ca availability, will be most important in determining how cells respond to stimulus or alterations in their environment. We have proposed earlier (5) that a reduction of the pool of extracellular Ca available for replenishing internal pools may progressively diminish the ability of cells to respond to external stimulus, eventually leading to dysfunction. Such a sequence might be the cause of observable deficiencies, and also occur during the senescence process.

CELLULAR RESPONSE

The traditional view of Ca as a nutrient impinges on its role in signal transduction principally in the elucidation of cellular distribution, and identification of Ca pools, their availability, and the mechanisms for transmembrane transport. There are two related aspects that I would like to highlight. While recognizing that there are a number of mechanisms which putatively involve Ca in regulating cellular responses, I want to concentrate on those involving phosphoinositides. The particular aspects are those which are peculiar to plants.

It is increasingly clear that between plants and animals there are significant differences in the operation of the phosphatidylinositol (PtdIns) phosphorylating cycle, and this may have strong implications for whether a system involving inositol trisphosphate (InsP3)-induced Ca release is important in plant cells. The most important differences are the relatively low levels of phosphatidylinositol bisphosphate (PtdInsP2) found in plant membranes, and the relatively high turnover of phosphatidylinositol monophosphate (PtdInsP) - or the high activity of the PtdIns phosphorylating enzyme (4, 11). Although Sommarin and Sandelius (18) have shown *in vitro* activity of PtdInsP kinase similar to that from animal tissues, incorporation of radiolabel into PtdInsP *in vivo* is initially high and rapid (4). Thus, the PtdIns phosphorylating cycle appears to be biased to the PtdIns end. The very low levels of PtdInsP2 raise the question of whether InsP3 can operate as in animal cells. It clearly is involved in pulvini responses (14), but other evidence is proving difficult to assemble. We should be asking then whether the phosphoinositide system exists in plants as in animal cells, and if different, how is this related to possible Ca responses.

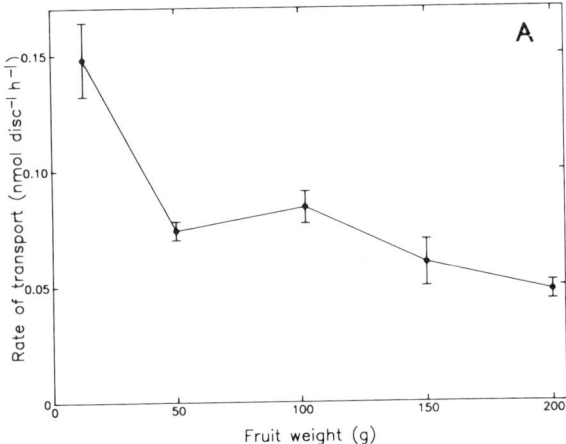

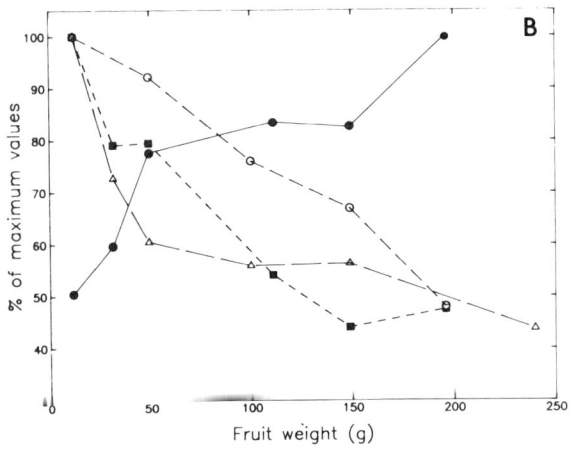

Figure 4. **A,** Rate of Ca transport across discs of the cortical flesh of apple fruit of increasing weight, taken at progressive stages of fruit development. Transport was measured by monitoring 45 Ca movement across discs from one solution-filled chamber to another. **B,** Changes in composition of apple fruit cortical tissue during fruit development. Air space (•); relative cell wall surface area (Δ); cation exchange capacity (■); cell wall yield (o). Figure reproduced from Harker and Ferguson (9) with permission of Physiologia Plantarum.

The other aspect of importance in Ca nutrition is the nature of releasable Ca pools. Transient rises in free Ca activity in the cytosol as a stimulus response might originate from voltage-sensitive channels in the plasma membrane, or from Ca release from internal pools. In animal cells, the ER is seen as the compartment providing releasable Ca, although recent work on the InsP3-binding protein has shown the latter's presence also in the plasma membrane (7). In plants, we have evidence, all *in vitro*, showing InsP3-induced release of Ca from crude microsomes (1, 3, 16), tonoplast vesicles (17) and whole vacuoles (15). The only experiment performed with purified ER (from red beet, 13), did not result in Ca release. While the ER and vacuole are potentially the most obvious sites, the ER has not been extensively explored in this regard, and there might be a problem in accepting that a comparatively large-volumed compartment such as the vacuole would be able to respond with rapid, often oscillatory, release. Reception and release would probably take place at specific sites on the tonoplast, raising the question of whether sufficient Ca would be available at the right place. However, recent work in animal cells, summarized by Cheek (2) shows that transient rises in cytosolic Ca can be very localized in the cell. This can be from localization of surface receptors, localization within the cell of the releasable Ca pool, and by clustering of Ca channels in the plasma membrane. In contemplating the vacuole as a usable Ca pool, we have to be able to account for the ensuing spatial problems.

A view of the involvement of Ca in plant cell response from the perspective of Ca as a nutrient inevitably highlights our need for more information on the importance of particular pools and sites of Ca in plant tissue. The modulation of the Ca content of these pools, particularly of the available Ca in the extracellular fluid, is the essence of a nutritional approach to the biochemistry and physiology of Ca in plants.

LITERATURE CITED

1. Canut H, Carrasco A, Graziana A, Boudet AM, Ranjeva R (1989) Inositol trisphosphate-stimulated calcium release from *Acer* microsomal fractions involves the uptake of potassium. FEBS Letts **253**: 173-177
2. Cheek TR (1989) Spatial aspects of calcium signalling. J Cell Sci 93: 211-216
3. Drobak BK, Ferguson IB (1985) Release of Ca^{2+} from plant hypocotyl microsomes by inositol-1,4,5-trisphosphate. Biochem Biophys Res Commun **130**: 1241-1246
4. Drobak BK, Ferguson IB, Dawson AP, Irvine RF (1988) Inositol-containing lipids in suspension-cultured plant cells. An isotopic study. Plant Physiol **87**: 217-222.
5. Ferguson IB, Drobak BK (1988) Calcium and the regulation of plant growth and senescence. HortScience **23**: 262-266
6. Ferguson IB, Watkins CB (1989) Bitter pit in apple fruit. Hort Rev **11**: 289-355
7. Furiuchi T, Yoshikawa S, Miyawaki A, Wada K, Maeda N, Mikoshiba K (1989) Primary structure and functional expression of the inositol 1,4,5-trisphosphate-binding protein P400. Nature **342**: 32-38

8. Hanson JB (1984) The functions of calcium in plant nutrition. Adv Plant Nutr **1**: 149-208
9. Harker FR, Ferguson IB (1988) Calcium ion transport across discs of the cortical flesh of apple fruit in relation to fruit development. Physiol Plant **74**: 695-700
10. Harker FR, Ferguson IB, Dromgoole FI (1988) Calcium ion transport through tissue discs of the cortical flesh of apple fruit. Physiol Plant **74**: 688-694
11. Irvine RF, Letcher AJ, Lander DJ, Drobak BK, Dawson AP, Musgrave A (1989) Phosphatidylinositol(4,5)bisphosphate and phosphatidylinositol(4)phosphate in plant tissues. Plant Physiol **89**: 888-892
12. Jones HG, Samuelson TJ (1983) Calcium uptake by developing apple fruits. II. The role of spur leaves. J Hort Sci **58**: 183-190.
13. Lew RR, Briskin DP, Wyse RE (1986) Ca^{2+} uptake by endoplasmic reticulum from *Zucchini* hypocotyls. The use of chlorotetracycline as a probe for Ca^{2+} uptake. Plant Physiol **82**: 47-53
14. Morse MJ, Crain RC, Satter RL (1987) Light-stimulated inositolphospholipid turnover in *Samanea saman* leaf pulvini. Proc Natl Acad Sci USA **84**: 7075-7078
15. Ranjeva R, Carrasco A, Boudet AM (1988) Inositol trisphosphate stimulates the release of calcium from intact vacuoles from *Acer* cells. FEBS Letts **230**: 137-141
16. Reddy ASN, Poovaiah BW (1987) Inositol 1,4,5-trisphosphate induced calcium release from corncoleoptile microsomes. J Biochem **101**: 569-573
17. Schumaker KS, Sze H (1987) Inositol 1,4,5-trisphosphate releases Ca^{2+} from vacuolar membrane vesicles of oat roots. J Biol Chem **262**: 3944-3946
18. Sommarin M, Sandelius AS (1988) Phosphatidylinositol and phosphatidylinositol-phosphate kinases in plant plasma membranes. Biochim Biophys Acta **958**: 268-278
19. Wallace A, Frolich E, Lunt OR (1966) Calcium requirements of higher plants. Nature **209**: 634

CALCIUM, CELL WALLS AND GROWTH

ROBERT E. CLELAND, SARBJIT S. VIRK, DOUGLAS TAYLOR
AND THOMAS BJÖRKMAN

*Department of Botany, KB-15, University of Washington,
Seattle, WA 98195, USA*

Most of the recent interest in calcium in plants has centered around its role in the cytoplasm in controlling developmental processes. However, more than half of the calcium in plants exists in the extracellular regions, *i.e.* in the apoplast. The possibility that this calcium might play a significant role in the control of plant growth should not be ignored.

One widely-considered role of calcium is that of providing wall rigidity by cross-linking the pectic chains. Bennett-Clark (1) suggested that auxin might be inducing cell enlargement by causing cleavage of these calcium crosslinks. One known effect of auxin is to cause cells to acidify their apoplast (3); since protons compete with calcium for pectic carboxyl groups, this acidification might cleave calcium bridges and thus cause wall loosening. A purpose of this investigation was to test this concept.

Free calcium in the apoplast may also influence plant growth. Exogenous calcium is an effective inhibitor of the growth of isolated stem sections. Sentenac and Grignon (10) have suggested that free calcium may modify the apoplastic pH by altering the Donnan Space cation composition. It has been difficult to test their ideas since neither the free calcium concentration nor the pH of the apoplast has been determined experimentally. A second objective of this investigation was to obtain these needed data.

MATERIALS AND METHODS

The plant material consisted of 15-mm sections cut 3 to 18 mm below the hook of soybean seedlings (*Glycine max* L., cv Williams 82) which had been grown for 3 to 4 d in dim red light (14). The sections were treated in one of three ways to compromise the cuticle barrier. Some were rubbed with Rottenstone (abraded) while others were bisected longitudinally (bisected). Finally, the outer 3 to 4 cell layers were removed from others by peeling with forceps (epidermal peels). Abraded and bisected sections were generally pretreated for 1 h in water before use.

For measurement of wall strength or of facilitated creep, bisected sections or epidermal peels were frozen on dry ice, thawed, and then pretreated as indicated before mechanical analysis. Wall strength was determined by use of the Instron technique, using the procedures described in Virk and Cleland (14). For facilitated creep assays, the thawed material was placed between the clamps of a constant-stress extension apparatus and placed under 10 to 20g load and in a pH 6.5 solution. After the viscoelastic extension had largely been completed (40 min), the solution was changed to one to be tested and the resulting change in section length was recorded over the next 6 h (2).

Wall calcium was determined by oven-drying, weighing, and then extracting sections with nitric acid. Calcium was then measured with an atomic absorption spectrophotometer and was expressed as μg calcium per gram dry weight of wall material (14).

To measure the equilibrium between free Ca^{2+}, bound Ca^{2+}, and free-space pH, frozen-thawed epidermal strips were incubated for 3 h in 20 mL of 20 mM Na-acetate buffers, pH 3.0 to 6.0, with 1 to 300 μM $CaCl_2$. The strips were then quickly rinsed with water, dried, and their weight and calcium content was determined.

Free-space pH was evaluated from the rate of uptake of the weak acid benzoic acid (10). Abraded sections were preincubated for 10 min in 50 μM benzoic acid, then transferred to a similar solution containing 20 nanocuries per mL ^{14}C-benzoic acid. After 15 min, the sections were rinsed for 2 min in distilled water, digested for 36 h in digesting fluid, and counted in a scintillation counter. The master pH curve was determined using 25 mM buffers between 3.0 and 7.0. The free-space pH was then evaluated by incubating sections in the absence of buffer and with or without 0.5 mM $CaCl_2$.

All experiments were repeated a minimum of twice, and each determination was performed in triplicate.

RESULTS AND DISCUSSION

Apoplastic Calcium Content

The cell walls of epidermal peels from the growing regions of soybean hypocotyls contain 800 to 1,000 μg Ca^{2+} per g dry weight, whether the walls are prepared by freezing-thawing or boiling in methanol (14). This value of 20 to 25 μmol/g dry weight is certainly an overestimate, since some vacuolar calcium is adsorbed onto the walls during isolation. Removal of protein from the walls does not lower the calcium content, indicating that the calcium is not bound to proteins.

If one assumes that the apoplast *in vivo* contains 2 g water per g dry weight (which is probably an overestimate), then one can calculate the amount of free calcium that would exist in the apoplast, assuming particular free calcium concentrations. If the free Ca^{2+} is 1 mM, the apoplast would contain 2 μmol/g dry weight; *i.e.* 10% of the calcium would be free and 90% bound. If the free Ca^{2+} is 0.1 mM, on the other hand, only 1% of the apoplastic calcium would be free and 99% bound. This relationship is important because it points out that solubilization of bound calcium may cause only a slight change in the amount of bound calcium, while causing a much larger change in the free Ca^{2+} level.

The Role of Wall-bound Calcium in Cell Enlargement

The possibility that calcium bridges between pectic chains are load-bearing bonds and that cleavage of these bridges might be a major mechanism of wall loosening has been frequently suggested (1, 12). Since auxin causes cells to acidify the apoplast and protons can solubilize wall-bound calcium, auxin-induced wall loosening may occur via cleavage of these calcium bridges. There are several predictions which can be used to test the validity of this idea. The first is that the mechanical strength of the walls, as evaluated by the Instron assay, should be reduced upon removal of calcium from the walls. Treatment of either methanol-boiled or frozen-thawed bisected sections or epidermal strips with H^+ (pH 3.5 solution) or with the calcium chelator EGTA resulted in a large increase in instron plastic (PEx), but only after the calcium content of the walls had been reduced to below about 5 μmol/g dry weight (14). Since EGTA acidifies the solution as it exchanges protons for bound calcium (8), this reduction in PEx might have been due to some unrelated, acid-mediated wall loosening. However, nearly identical results were obtained with the calcium chelator 2-((2-bis-(carboxymethyl)amino-5-methylphenoxy)-methyl)-6-methoxy-8-bis(carboxymethyl)aminoquinoline (Quin-2), which does not change the wall pH. These data suggest that calcium bridges can be load-bearing bonds, but that wall loosening does not occur until over two-thirds of the wall calcium is solubilized.

A second prediction is that if frozen-thawed walls are placed under constant tension, cleavage of calcium bridges with Quin-2 will cause extension comparable to that caused by acidic solutions (facilitated creep). Quin-2, at 1 mM, removed wall calcium almost completely within the first hour. There was a small amount of facilitated creep which occurred in response to Quin-2, but its magnitude was nowhere near as great as that induced by a pH 3.5 solution (Table I).

Table I. *Removal of Wall Calcium by Quin-2 Does Not Result in Facilitated Creep*

Frozen-thawed bisected sections were extended under 20g force in 50 mM Na-acetate buffer, pH 6.0 or 3.5, ± 1 mM Quin-2. Extension rate after 2 h and calcium remaining in wall at end of extension are given.

Conditions	Creep rate	Wall calcium
	μm/h	μg Ca^{2+}/g dry weight
pH 6.0, no Quin-2	0.21±0.03	575
pH 6.0 + Quin-2	0.66±0.45	17
pH 3.5, no Quin-2	1.58±0.28	125

A final indication that cleavage of calcium bridges by protons is not the mechanism of auxin-induced wall loosening comes from a consideration of the amount of bound calcium solubilized by a pH 4.5 solution--a pH known to occur in the walls of auxin-treated tissues (3). When frozen-thawed epidermal strips were incubated for 4 h in a pH 4.5 Na-acetate buffer, only about 20% of the initial wall calcium was removed. This is not enough to cause an effective increase in PEx or any facilitated creep. We are forced to conclude that calcium bridges are not the major load-bearing bonds in soybean hypocotyl cell walls, and that H^+-induced wall loosening does not occur via cleavage of these calcium bridges.

The Role of Free Calcium in the Control of Cell Elongation

The physiological role of free apoplastic calcium has received scant attention. One reason for this is the uncertainty as to the actual concentration of free Ca^{2+}. Estimates range from 10 μM to 10 mM (5, 7), but actual measurements hardly exist. We have used calcium microelectrodes to obtain the free Ca^{2+} concentration. A thin (<0.5 mm) longitudinal epidermal strip was removed from the growing zone of intact soybean hypocotyls and measurements were made immediately. Readings of pCa for a particular hypocotyl were repeatable, both in time and along the length of the strip. However, readings varied considerably from one hypocotyl to another. Eleven readings ranged from 0.05 to 0.57 mM Ca^{2+}, although most of the values fell between 0.10 and 0.22 mM. The average of the 11 readings was 0.17 mM.

In the apoplast *in vivo*, there should be an equilibrium between the amount of bound calcium, the free Ca^{2+}, and the H^+ concentration. To determine this relationship, we incubated frozen-thawed epidermal strips in solutions containing 20 mM Na-acetate buffers, pH 3.0 to 6.0 and with Ca^{2+} concentrations between 1 and 300 μM. After 3 h, the strips were carefully rinsed and their content of bound calcium was determined. A solution in which the strip neither gained nor lost calcium, relative to the initial value, can be assumed to have had a pCa and pH that might have existed in the apoplast *in vivo* in equilibrium with the bound calcium. When the possible values for H^+ *versus* Ca^{2+} for such possible equilibrium conditions were plotted, a straight line was obtained for Ca^{2+} values between 5 and 300 μM and for pHs between 3.0 and 4.5 (Fig. 1). By assuming or knowing the apoplastic Ca^{2+} concentration, one can predict the pH that must have existed in the wall. Likewise, assuming a particular wall pH, one can predict the free Ca^{2+} concentration. If the wall pH is about 5.0 (3), the free Ca^{2+} must be only 1 to 3 μM, far below the measured values. If the free Ca^{2+} is 100 μM, then the wall pH must be 3.5 or below. Which of these situations is correct?

The apoplast, with its large concentration of fixed cell wall anions (COO^-), consists of both a Free Space (FS) and a Donnan Free Space (DFS). The DFS will contain an excess of cations compared with the FS, in order to provide charge balance. These cations will be in solution, but will be effectively "trapped." The various cations differ in their ability to be DFS cations;

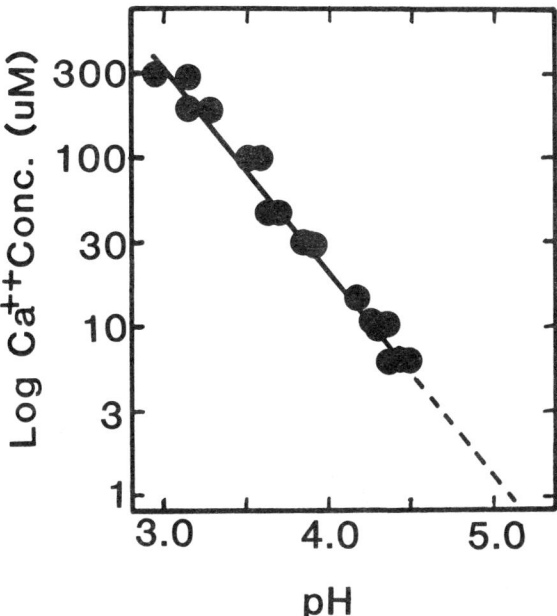

Figure 1. Conditions of free Ca^{2+} and H^+ that would be in equilibrium with the *in vivo* wall-bound Ca^{2+}. Frozen-thawed epidermal peels were incubated in 1-300 μM Ca^{2+} at pHs between 3.0 and 6.0. For each Ca^{2+} level, a pH was determined at which calcium was neither lost nor gained from the walls. This relationship predicts that if the free Ca^{2+} was 100 μM, the wall pH would have to be about 3.5.

$H^+ > Ca^{2+} >> K^+ > Na^+$. The H^+ which are DFS cations will cause the DFS to have a pH lower than that of the FS. Sentenac and Grignon (10) have suggested that the DFS can be as much as 2.5 pH units lower than the FS. Since the pH that determines the H^+ *versus* free Ca^{2+} *versus* bound calcium equilibrium is the pH at the wall carboxyl groups, it must be the DFS pH that influences this equilibrium. A DFS pH of 3.5 when the FS pH is 5.0 is not unreasonable.

This difference in pH between DFS and FS may explain how exogenous calcium inhibits growth. It is well known that exogenous calcium causes rapid inhibition of auxin-induced growth of isolated stem sections. The characteristics of this inhibition in abraded soybean hypocotyl sections are surprising. First of all, a concentration of only 0.1 mM Ca^{2+} inhibits the growth of abraded, auxin-treated soybean hypocotyl sections by 50% within the first 30 min (13), yet this level of calcium does not inhibit auxin-induced proton excretion, acid-induced wall loosening or cell wall synthesis. Secondly, the inhibition is transitory. The length of time that growth is inhibited depends on the Ca^{2+} concentration; with 0.5 mM Ca^{2+}, the inhibition persists for only about 30 min, while with 1 mM it persists for over 1 h. This can be understood by reference to the effect of exogenous Ca^{2+} on the DFS pH. The added calcium ions will permit protons to leave the DFS for the FS. The resulting increase in DFS pH will inhibit wall loosening enzymes which possess acidic pH optima. This disequilibrium cannot persist, however, and with time the original DFS and FS pHs will be restored and growth will resume.

Direct evidence for such a mode of inhibition will require a measurement of the DFS pH. Studies using ^{13}C-NMR to measure the actual carboxyl pH are underway, but not yet completed. An indirect approach is to determine the effect of added Ca^{2+} on the FS pH, since an increase in DFS pH would have to be accompanied by a decrease in FS pH. Use has been made of the fact that weak acids, such as benzoic acid, only enter cells as the undissociated molecule, and the proportion of benzoic acid existing in the undissociated form, and thus taken up into the cell depends on the pH next to the plasma membrane. By measuring the uptake rate of benzoic acid into abraded sections over short times and at varying pHs, one can obtain a master pH curve for uptake. The uptake rate for benzoic acid in water indicates that the FS pH in the absence of auxin is about 5.8 (13). Addition of 0.5 mM Ca^{2+} results in a reduction in FS pH by about one pH unit. This is the expected result if the DFS pH increases in response to exogenous calcium.

A calcium-induced change in the DFS pH may also form an integral part of the auxin-induced growth mechanism. Auxin-induced growth of soybean hypocotyl sections requires an acidic apoplast, as shown by the ability of neutral buffers to prevent wall loosening and growth (9). The enzymes with an acidic pH optimum, responsible for the wall loosening, are presumedly in

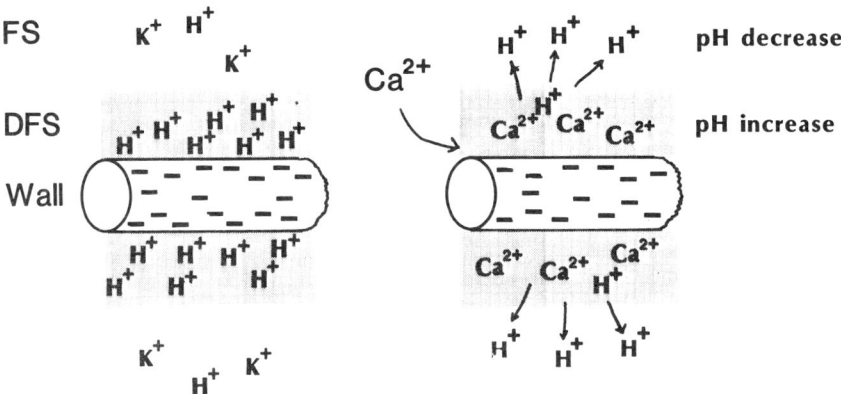

Figure 2. Effect of exogenous Ca^{2+} on the DFS and FS pH. In the absence of Ca^{2+}, H^+ comprises the bulk of the DFS cations, making the DFS pH < FS pH. Added Ca^{2+} replaces H^+ as the DFS cation, allowing H^+ to move to the FS. The result is an increase in DFS pH and a decrease in FS pH.

the DFS. If auxin were to open calcium channels in the plasma membrane and permit Ca^{2+} uptake into the cytoplasm, the FS would be depleted of free Ca^{2+}. This, in turn would lead to a loss of free Ca^{2+} from the DFS in exchange for H^+ from the FS. The result would be acidification of the DFS and, thus, enhanced wall loosening without a decrease in FS pH. In those systems such as the lettuce hypocotyl, where no GA-induced proton excretion occurs, uptake of apoplastic calcium has been implicated in the growth mechanism (6). Such an idea would be consistent with the observation that cytoplasmic Ca^{2+} increases in maize coleoptiles within minutes of the addition of auxin (4).

CONCLUSIONS

1) Calcium crosslinks are not major load-bearing bonds in soybean hypocotyl cell walls. Protons cause wall loosening by some mechanism other than displacement of wall calcium.

2) The apoplastic free Ca^{2+} is in the 100 to 200 μM range. The pH of the DFS is predicted to be about 3.5 in soybean hypocotyls.

3) Exogenous calcium inhibits growth by releasing H^+ from the DFS to the FS, raising the pH of the DFS and inhibiting wall loosening enzymes with acidic pH optima.

ACKNOWLEDGEMENTS

This research was supported by grants from NASA and from the Department of Energy.

LITERATURE CITED

1. **Bennett-Clark TA** (1956) Salt accumulation and mode of action of auxin. In RL Wain, F Wightman, eds, The Chemistry and Mode of Action of Plant Growth Substances. Butterworths, London, pp 284-291
2. **Cleland RE, Cosgrove D, Tepfer M** (1987) Long-term acid-induced wall extension in an in-vitro system. Planta **170**: 379-385
3. **Cleland RE, Rayle DL** (1978) Auxin, H^+-excretion and cell elongation. Bot Mag Tokyo, Spec Iss **1**: 125-139
4. **Felle H** (1988) Auxin causes oscillations of cytosolic free calcium and pH in Zea mays coleoptiles. Planta **174**: 495-499
5. **Macklon AES** (1975) Cortical cell fluxes and transport to the stele in excised root segments of Allium cepa L. II. Calcium. Planta **122**: 131-141
6. **Moll C, Jones RL** (1981) Calcium and GA-induced elongation of lettuce hypocotyl sections. Planta **152**: 450-456
7. **Raven JA** (1985) Long distance transport of calcium. In AJ Trewavas, ed, Molecular and Cellular Aspects of Calcium in Plant Development. Plenum Press, NY, pp 241-250
8. **Rayle DL** (1989) Calcium bridges are not load-bearing cell-wall bonds in Avena coleoptiles. Planta **178**: 92-95
9. **Rayle DL, Cleland RE** (1980) Evidence that auxin-induced growth of soybean hypocotyls involves proton excretion. Plant Physiol **66**: 433-437

10. **Sentenac H, Grignon C** (1981) A model for predicting ionic equilibrium concentration in cell walls. Plant Physiol **68**: 415-419
11. **Sentenac H, Grignon C** (1987) Effect of H^+ excretion on the surface pH of corn root cells evaluated by using weak acid influx as a pH probe. Plant Physiol **84**: 1367-1372
12. **Soll HJ, Böttger M** (1982) The mechanism of proton-induced increase in wall extensibility. Plant Sci Lett **24**: 163-171
13. **Taylor D** (1990) The role of calcium in the regulation of elongation growth in the soybean hypocotyl. PhD Thesis. University of Washington, Seattle, WA
14. **Virk SS, Cleland RE** (1988) Calcium and the mechanical properties of soybean hypocotyl cell walls: possible role of calcium and protons in wall loosening. Planta **176**: 60-67

CALCIUM BINDING BY CHLOROPLAST STROMA PROTEINS AND FUNCTIONAL IMPLICATIONS

Maria Burchert, Barbara Surek, Georg Kreimer, and Erwin Latzko

*Botanisches Institut der Westfaelischen Wilhlems-Universität,
Schlossgarten 3, D-4400 Münster, FRG*

In plant cells, most of the calcium content is localized in the apoplast. For chloroplasts, calcium levels between 4 and 23 mM have been reported (17). The levels depend on the age of the tissue (1). The large variation in Ca^{2+} levels may also depend on the isolation procedure. EDTA present in the media used for isolation of chloroplasts decreases the calcium level nearly 50% (11). The high level of calcium and its further increase upon illumination present a paradox because CO_2-assimilation by isolated chloroplasts is already inhibited 50% at 0.5 mM Ca^{2+} in the assay medium (2). Possible targets for calcium inhibition of CO_2-fixation are the chloroplast fructose and sedoheptulose bisphosphatases. The catalytic activity of these enzymes is completely inhibited at Ca^{2+} concentrations of approximately 0.3 mM. Fructose-1,6-bisphosphatase [EC 3.1.3.1] (FBPase) has an $I_{0.5}$ for Ca^{2+} of 7 to 40 µM (6), and sedoheptulase-1,7-bisphosphatase [EC 3.3.3.37] (SBPase) has an $I_{0.5}$ for Ca^{2+} of 10 to 60 µM (21).

Therefore, we have to assume that only a minor portion of the total calcium in the chloroplast is metabolically active as free Ca^{2+} and that the major portion of this ion is firmly bound. Several ways of achieving a reduced free Ca^{2+} level have to be considered: (*i*) binding to membranes and membrane constituents, and (*ii*) binding to stroma proteins and nonprotein low-mol-wt components.

We have estimated both the binding capacity and the affinity of stroma proteins for calcium (12). Stroma fractions were prepared from isolated spinach chloroplasts (3), and the associated Ca^{2+} was removed by passing a concentrated fraction through Sephadex G25/Chelex. Binding of Ca^{2+} was measured using the metallochromic indicator tetramethylmurexide (16). From the change in differential absorbance (507/544 nm), free and bound calcium can be calculated using the K_D of the Ca^{2+} indicator complex (12).

The total number of binding sites (n_o) of stroma proteins varied between 90 and 155 nmol · mg^{-1} protein, with average binding constants ($<k_o>$) between 370 and 909 µM. Both Mg^{2+} and La^{3+} were found to inhibit Ca^{2+} binding competitively, whereas K^+ had no effect on either the binding sites or the binding constants ($I_{0.5}$ Mg^{2+} = 3.85 mM; $I_{0.5}$ La^{3+} = 25 µM).

In order to quantify the binding capacity of the stroma for calcium that might occur during dark/light transition, binding experiments were conducted at pH 7.1 and 7.8 to mimic the pH prevailing in the dark-kept and illuminated chloroplasts. At pH 7.1, the number of Ca^{2+}-binding sites decreased from 90 to 59 nmol · mg^{-1} protein. However, the average affinity for these sites increased at pH 7.1 from 370 to 220 μM.

Finally, we have asked how effective is the capacity of the chloroplast to buffer the large variation of Ca^{2+} levels and to maintain and effectively control the level of free Ca^{2+} in the stroma within the micromolar range.

The results of these calculations are presented in Figure 1. These calculations are based on the binding constants and number of binding sites for the thylakoids as obtained by Gross and Hess (4). For the stroma proteins, our data were used (12).

The theoretical relationship between bound and free calcium indicates that the free Ca^{2+} concentration in the chloroplast can be maintained below 10 μM up to a total Ca^{2+} concentration of about 18 mM at pH 7.8 (light conditions). At pH 7.1 (dark conditions), the free Ca^{2+} concentration is always higher than at pH 7.8 (12). These values are consistent with our measured data for the stromal-free Ca^{2+} concentration in dark-kept chloroplasts (10). The buffer capacity of the chloroplast seems to be strong enough to control the free Ca^{2+} concentration, even up to a total Ca^{2+} concentration of about 20 to 24 mM.

For a direct determination of the free Ca^{2+} level in the stroma, we have applied a Ca^{2+} null point titration technique using dark-kept chloroplasts (10). Titrations were performed in the presence of the ionophore A 23187, which catalyzes an electroneutral $H^+/metal^{2+}$ exchange (Fig. 2). Thus, when the extrachloroplastic-free Ca^{2+} concentration is such that no net movement of calcium across the chloroplast envelope occurs upon addition of the inophore

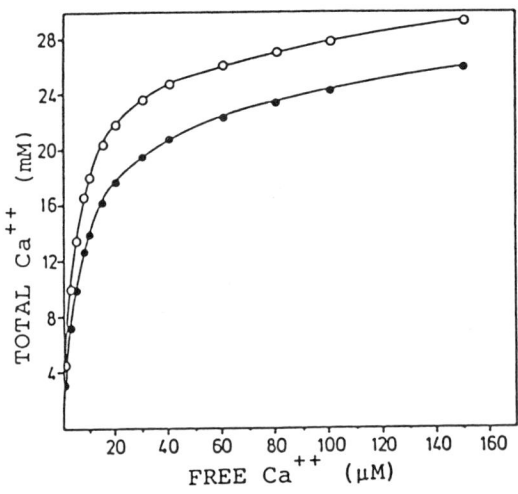

Figure 1. Theoretical relationship between the total Ca^{2+} content of chloroplasts and the free Ca^{2+} concentration in the stromal space.
o = pH 7.8, 6 mM Mg^{2+};
● = pH 7.1, 3 mM Mg^{2+}.

A 23187 (= Ca^{2+} null point = intersection with the abscissa), the external and internal concentrations of Ca^{2+} and H^+ will be related to the equation:

$$[Ca^{2+}]_{out}/[Ca^{2+}]_{in} = 10^{2\,\Delta\,pH}$$

The stromal concentration of free Ca^{2+} can be calculated if the proton gradient is known. Therefore, we have determined the Δ pH across the envelope under the conditions of the Ca^{2+} null point titration. Since no Δ pH was measurable under these conditions, the stromal concentration of free Ca^{2+} equals the Ca^{2+} null point. The values for free Ca^{2+} vary between 2.4 and 6.3 μM, as obtained for chloroplasts from 12 different samples of spinach leaves. Comparisons with osmotically shocked chloroplasts confirmed the Ca^{2+} null point measured with intact chloroplasts. The Ca^{2+} null point of osmotically shocked chloroplasts was 17.5 μM, which was higher than that of intact chloroplasts (Fig. 2).

Until recently, specific binding of Ca^{2+} has only been proven for chloroplast FBPase (7). However, since a large proportion of stromal proteins exhibit acidic isoelectric points, it is very likely that further proteins have to be considered as Ca^{2+}-binding entities. One of these proteins is ferredoxin (20). Further, likely candidates are the chloroplast thioredoxins f and m. Therefore, we have examined Ca^{2+} binding by purified samples of these proteins using ^{45}Ca autoradiography on nitrocellulose membranes (13).

As shown in Figure 3a, reference proteins and ferredoxin were blotted onto nitrocellulose after SDS-PAGE and stained by amido black. The corresponding autoradiograph of the ^{45}Ca-incubated nitrocellulose is demonstrated in Figure 3b.

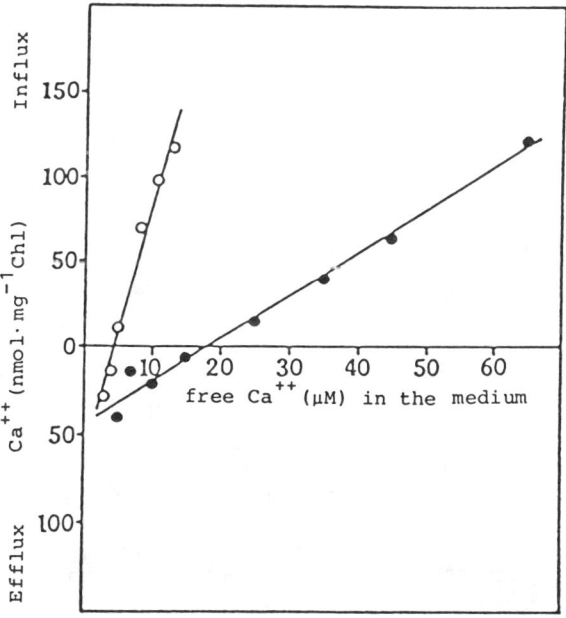

Figure 2. Comparison of Ca^{2+} null points from intact and osmotically shocked chloroplasts. Intact chloroplasts (o) and osmotically shocked chloroplasts (●).

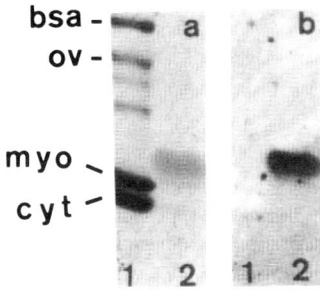

Figure 3 a,b. Ferredoxin and protein standards after SDS-PAGE (15% gel) and electrophoretic transfer to nitrocellulose. a, Nitrocellulose membrane, showing the protein bands stained by amido black (after autoradiography). b, Autoradiograph of the ^{45}Ca-incubated nitrocellulose membrane. Lane 1: protein standards (bsa, bovine serum albumine; ov, ovalbumine; myo, myoglobin; cyt, cytochrome c. Lane 2: purified ferredoxin. pH of the ^{45}Ca-overlay buffer = pH 6.8; MgCl$_2$ concentration = 5 mM.

The results of ^{45}Ca autoradiography of denatured thioredoxins are presented in Figures 4a and b. Ferredoxin and thioredoxins appear heavily labeled, while the protein standards did not bind ^{45}Ca. Labeling was obtained with both denatured and native proteins (Fig. 5).

The binding of Ca^{2+} to ferredoxin and thioredoxins is characterized by a high degree of specificity. This is demonstrated by the reduction of Ca^{2+} binding observed in the presence of micromolar concentrations of the Ca^{2+} antagonist LaCl$_3$ (data not shown). In the presence of 0.8 µM Ca^{2+}, only millimolar concentrations of MgCl$_2$ caused a similar effect. Further investigations showed that both oxidized and reduced ferredoxin and thioredoxins bind Ca^{2+}. In the case of ferredoxin, the reduced protein binds between 30 and 40% more radioactivity than the oxidized protein. Similar results were obtained with reduced and oxidized thioredoxins (data not shown).

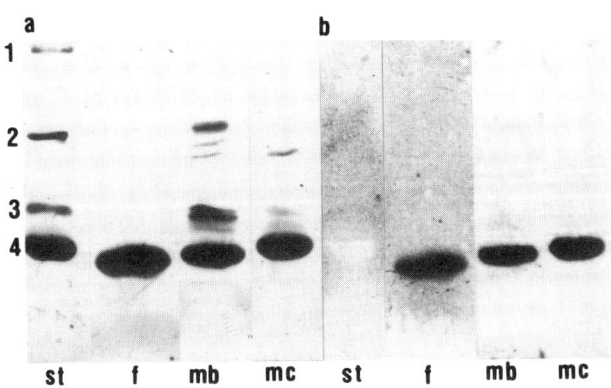

Figure 4 a,b. Thioredoxins and protein standards after SDS-PAGE (15% gel) and electrophoretic transfer to nitrocellulose. a and b as in Figure 3. Standard proteins: 1, bovine serum albumine; 2, carboanhydrase; 3, β-lactoglobulin; 4, cytochrome c; f, thioredoxin f; mb, thioredoxin mb; mc, thioredoxin mc. The asterics indicate thioredoxin mb and mc, respectively. ^{45}Ca-overlay buffer was 10 mM Hepes pH 7.1 supplemented with 1 mM MgCl$_2$ and 60 mM KCl.

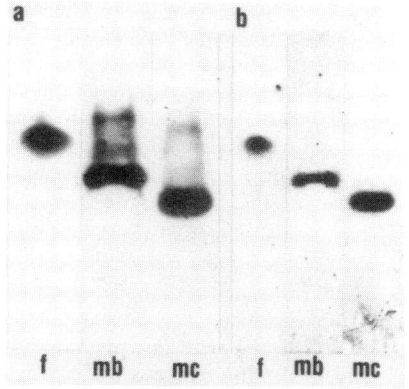

Figure 5 a,b. Thioredoxins after native PAGE (9% gel, Maurer I system) and electrophoretic transfer to nitrocellulose. a and b as in Figure 3. f, thioredoxin f; mb, thioredoxin mb; mc, thioredoxin mc. The asterics indicate thioredoxin mb and mc, respectively. The Ca-overlay buffer was as in Figure 4.

In the chloroplast, the phosphorylation of NAD yielding NADP by NAD kinase is responsible for the elevation of the NADP level upon illumination (15). The bulk of NAD kinase is located in the chloroplast. It is still controversial as to whether or not the chloroplast NAD kinase is regulated by Ca^{2+} and/or the Ca^{2+}-calmodulin complex (8, 19). In this respect, it may be relevant that no calmodulin-binding proteins have been detected in the stroma of spinach and pea chloroplasts (18).

Muto (14) suggested that light-induced Ca^{2+} influx into wheat chloroplasts may play a role in the regulation of NAD kinase activity. In order to investigate this possibility, we have used isolated spinach chloroplasts and have measured the effect of light-induced Ca^{2+} uptake on the NAD level (9). In control experiments, the light-induced Ca^{2+} uptake was inhibited by ruthenium red. A significantly lower decrease in the NAD level was observed if the Ca^{2+} influx was blocked (Fig. 6). As the NAD level is considerably larger than that of NADP (5, 15), even small-percentage decreases in NAD levels cause a strong increase of the NADP pool size.

However, no evidence is available so far as to whether Ca^{2+} is involved in the regulation of chloroplast FBPase *in vivo*. Therefore, we have determined the effect of light-induced Ca^{2+} influx on light-mediated activation of FBPase in the presence and absence of ruthenium red. The most

Figure 6. Effect of ruthenium red on light-induced changes in the NAD level of intact spinach chloroplasts in the presence of 50 μM Ca^{2+}. o, control; ●, +20 μM ruthenium red.

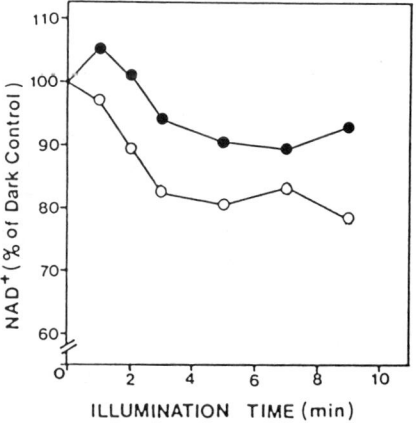

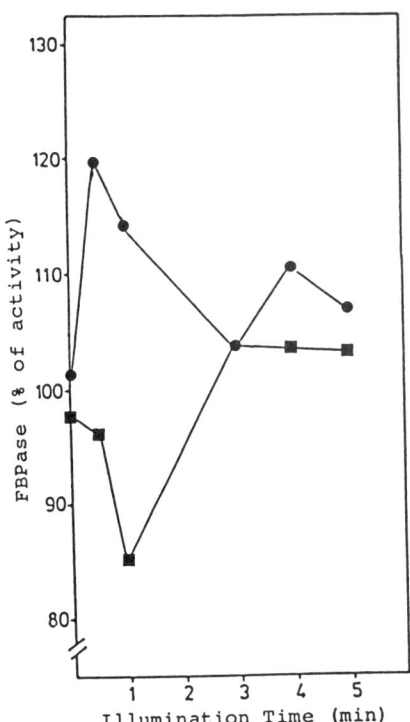

Figure 7. Effect of ruthenium red on the light-mediated activation of FBPase in chloroplasts. ●: Chloroplasts were incubated at 10^{-12} M and 10^{-5} M free Ca^{2+}. The activities at the indicated times were expressed as percentage of the respective activities at 10^{-12} M free Ca^{2+}. ■: Chloroplasts were incubated at 10^{-5} M free Ca^{2+} in the absence and presence of 20 μM ruthenium red. FBPase activity in the presence of ruthenium red is expressed as percentage of the activity without ruthenium red.

pronounced effect of Ca^{2+} on the activation of FBPase was observed during the first minutes of illumination (Fig. 7). After 5 min, however, the activation reached a similar plateau under both sets of conditions. We assume that either the level of free Ca^{2+} in the stroma is transiently reduced by illumination of the chloroplasts, or the FBPase is not yet saturated with Ca^{2+} at the onset of illumination.

In order to study the effect of Ca^{2+} on the activation of chloroplast FBPase in more detail, we have used stromal preparations. Since the catalytic activity of FBPase is strongly inhibited by Ca^{2+}, we had to preincubate the enzyme with increasing concentrations of Ca^{2+}, keeping the level of Mg^{2+} saturated (30 min). Prior to the assay of activity, the Ca^{2+} levels of the incubation media were adjusted with EGTA uniformly to 60 μM. The results clearly show that there is an essential demand for Ca^{2+} in order to activate chloroplast FBPase (Fig. 8). Maximal activation was achieved with 340 to 510 μM free Ca^{2+} using different stroma preparations.

For the direct demonstration of the effect of Ca^{2+} on the activation of chloroplast FBPase, we have preincubated the enzyme at 510 μM Ca^{2+} for 10 min and have injected samples into a reaction medium containing Ca^{2+}/EGTA to adjust the level of free Ca^{2+} to 60 μM. The activity was followed at increasing concentrations of DTE. The results are presented in Figure 9. Very little activation by DTE was obtained if Ca^{2+} was not present in the preincubation medium. From these experiments, we conclude that activation of chloroplast FBPase by reducing agents such as DTE or by the thioredoxin system strictly depends on the saturation of the Ca^{2+}-binding sites of the enzyme with Ca^{2+}.

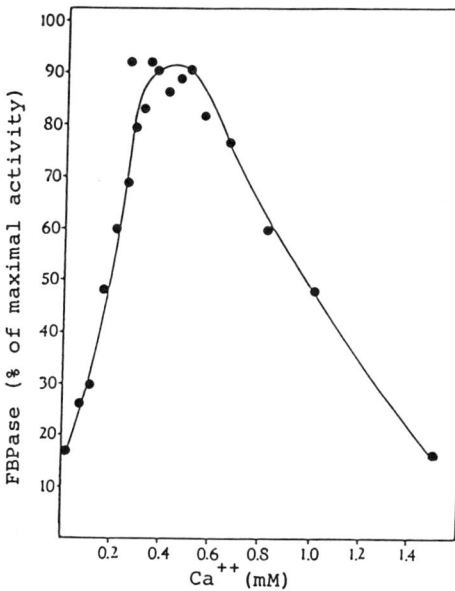

Figure 8. Effect of Ca^{2+} on the activation of FBPase. Stromal preparations (40-90 µg protein) were preincubated 30 min in the presence of the indicated concentrations of free Ca^{2+}, than injected into an assay medium where the Ca^{2+} concentration was adjusted uniformly to 60 µM by Ca^{2+}/EGTA.

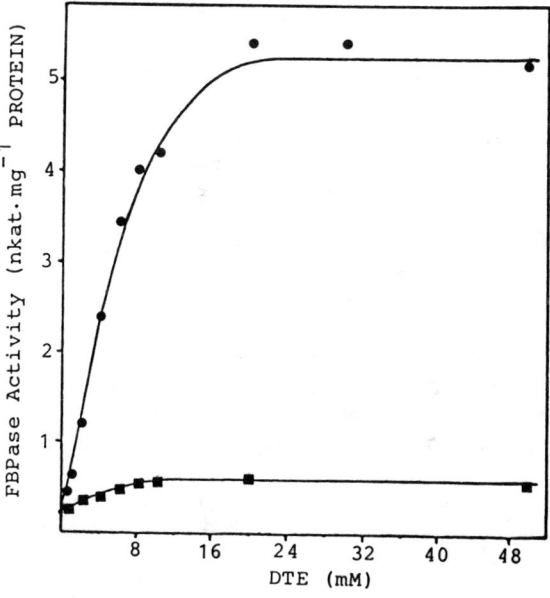

Figure 9. Effect of Ca^{2+} on the activation of FBPase by DTE. Stromal preparations (100 µg protein) were preincubated with 510 µM free Ca^{2+} (●) or without Ca^{2+} (■) and the indicated DTE concentrations were present in the assay medium.

Acknowledgments

Supported by the Deutsche Forschungsgemeinschaft: Grant LA 161.

LITERATURE CITED

1. **Bouthyette PY, Jagendorf AT** (1981) Calcium inhibition of amino acid incorporation by pea chloroplasts and the question of loss of activity with age. In G Akoyunoglou, ed, Photosynthesis, Vol 5, Chloroplast Development. Balaban International Sciences Service, Philadelphia, pp 599-609
2. **Demmig B, Gimmler H** (1979) Effect of divalent cations on cation fluxes across the chloroplast envelope and on photosynthesis of intact chloroplasts. Z Naturforsch **34c:** 233-241
3. **Douce R, Joyard J** (1982) Purification of the chloroplast envelope. In M Edelman, RB Hallick, NH Chua, eds, Methods in Chloroplast Molecular Biology. Elsevier Biomedical, Amsterdam, NY, Oxford, pp 239-256
4. **Gross EL, Hess SC** (1974) Correlation between calcium ion binding to chloroplast membranes and divalent cation-induced structural changes and changes in chlorophyll a fluorescence. Biochim Biophys Acta **339:** 334-346
5. **Heber U, Santarius KA** (1965) Compartmentation and reduction of pyridine nucleotides in relation to photosynthesis. Biochim Biophys Acta **109:** 390-408
6. **Hertig C, Wolosiuk RA** (1980) A dual effect of Ca^{2+} on chloroplast fructose-1,6-bisphosphatase. Biochem Biophys Res Commun **97:** 325-333
7. **Hertig CM, Wolosiuk RA** (1983) Studies on the hysteretic properties of chloroplast fructose-1,6-bisphosphatase. J Biol Chem **258:** 984-989
8. **Jarrett HW, Brown CJ, Black CC Jr, Cormier MJ** (1982) Evidence that calmodulin is in the chloroplast of peas and serves a regulatory role in photosynthesis. J Biol Chem **257:** 13795-13804
9. **Kreimer G, Melkonian M, Holtum JAM, Latzko E** (1985) Characterization of calcium fluxes across the envelope of intact spinach chloroplasts. Planta **166:** 515-523
10. **Kreimer G, Melkonian M, Holtum JAM, Latzko E** (1988) Stromal free calcium concentration and light-mediated activation of chloroplast fructose-1,6-bisphosphatase. Plant Physiol **86:** 423-428
11. **Kreimer G, Surek B, Heimann K, Burchert M, Lukow L, Holtum JAM, Woodrow IE, Melkonian M, Latzko E** (1987) Calcium metabolism in chloroplast and protoplasts. In J Biggins, ed, Progress in Photosynthesis Research, Vol 3. Martinus Nijhoff Publishers Dordrecht, Boston, Lancaster, pp 4.345-4.357
12. **Kreimer G, Surek B, Woodrow IE, Latzko E** (1987) Calcium binding by spinach stromal proteins. Planta **171:** 259-265
13. **Maruyama K, Mikawa T, Ebashi S** (1984) Detection of calcium binding proteins by ^{45}Ca autoradiography on nitrocellulose membrane after sodium dodecyl sulfate gel electrophoresis. J Biochem **95:** 511-519
14. **Muto S** (1982) Distribution of calmodulin within wheat leaf cells. FEBS Lett **147:** 161-164
15. **Muto S, Miyachi S, Usada H, Edwards GE, Bassham JA** (1981) Light-induced conversion of nicotinamide adenine dinucleotide to nicotinamide adenine dinucleotide phosphate in higher plant leaves. Plant Physiol **68:** 324-328
16. **Ogawa Y, Tanokura M** (1984) Calcium binding to calmodulin: effects of ionic strength, Mg^{2+}, pH and temperature. J Biochem **95:** 19-28

17. **Portis AR Jr, Heldt HW** (1976) Light-dependent changes of the Mg^{2+} concentration in the stroma in relation to the Mg^{2+} dependency of CO_2 fixation in intact chloroplasts. Biochim Biophys Acta **449**: 434-446
18. **Roberts DM, Zielinski RE, Schleicher M, Watterson DM** (1983) Analysis of suborganellar fractions from spinach and pea chloroplasts for calmodulin-binding proteins. J Cell Biol **97**: 1644-1647
19. **Simon P, Bonzon M, Greppin H, Marme D** (1984) Subchloroplastic localization of NAD kinase activity: evidence for a Ca^{2+}, calmodulin-dependent activity at the envelope and for a Ca^{2+}, calmodulin independent activity in the stroma of pea chloroplasts. FEBS Lett **167**: 332-338
20. **Surek B, Kreimer G, Melkonian M, Latzko E** (1987) Spinach ferredoxin is a calcium binding protein. Planta **171**: 565-568
21. **Wolosiuk RA, Hertig CM, Nishizawa AN, Buchanan BB** (1982) Enzyme regulation in C_4 photosynthesis. Role of Ca^{2+} in thioredoxin-linked activation of sedoheptulose bisphosphatase from corn leaves. FEBS Lett **140**: 31-35

CALCIUM, SALINITY AND THE PLASMA MEMBRANE

ANDRÉ LÄUCHLI

Department of Land, Air and Water Resources, University of California Davis, CA 95616, USA

Nonhalophytes are plants that are sensitive to saline environments. They suffer from various metabolic disorders, and inhibition of growth and development when subjected to saline conditions (17). High Na^+ concentrations can cause disturbances in Ca nutrition (24). Since the 1960s, several studies have demonstrated that many nonhalophytes are particularly sensitive to high Na^+/Ca^{2+} ratios in the medium (12, 17). On the other hand, supplemental Ca^{2+} can mitigate the detrimental effects of high Na^+ on growth (*e.g.* 7, 14). It is now well established that inadequate Ca^{2+} concentrations at high Na^+ adversely affect growth and function of membranes (18). However, the mechanisms by which Ca^{2+} exerts a protective effect on membranes under salt stress are still debated. In this paper, the main lines of research will be reviewed regarding the role of Ca^{2+} in the regulation of plant responses to salinity, and a cellular model will be presented which ties the results together.

GROWTH RESPONSE

Ben-Hayyim and Kochba (2) obtained salt-tolerant lines of *Citrus sinensis* from ovular callus which were maintained in the presence of 200 mM NaCl. Growth of such a salt-tolerant line under salt stress was clearly dependent on the Ca^{2+} supply (1). This is shown in Figure 1, which demonstrates that the higher the Na^+ concentration in the medium, the greater is the Ca^{2+} requirement for maximum growth. This result establishes a quantitative relationship between the nutritional Ca requirement and the degree of salt stress.

POTASSIUM/SODIUM SELECTIVITY

Maintenance of adequate K^+ status, and K^+/Na^+ selectivity in the plant, is necessary for normal cell functions under saline conditions (12). Ca is essential for the maintenance of K^+ transport (10) and K^+/Na^+ selectivity under salt stress (9). In the latter report, it was shown that high Ca^{2+} inhibited Na^+ influx, improved the Ca-status, and maintained K^+/Na^+ selectivity in the root of salt-stressed cotton seedlings. We are now studying the relationships

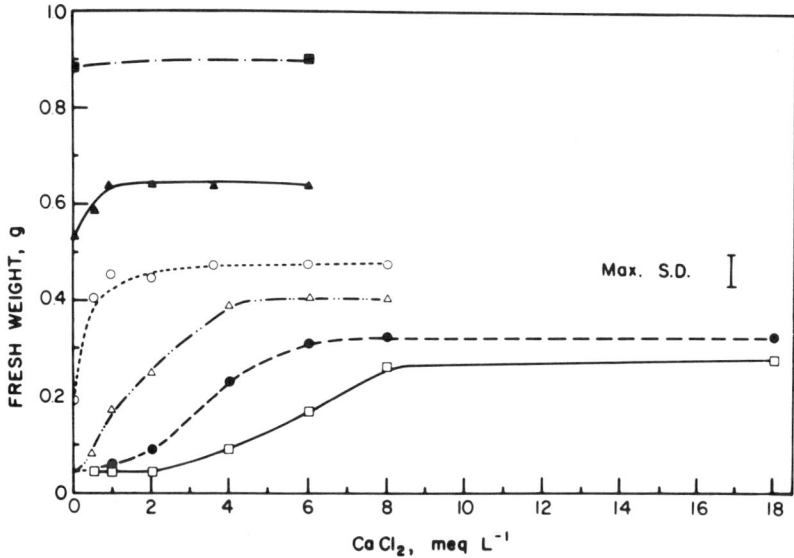

Figure 1. Effect of Ca^{2+} supply on growth of salt-tolerant *Citrus* cells at various Na^+ concentrations. ■, 0; ▲, 100; ○, 150; △, 200; ●, 250; □, 300 mM NaCl. Reproduced from (1) with permission of American Society of Plant Physiologists.

between maintenance of K^+/Na^+ selectivity by supplemental Ca^{2+} and root development by determining longitudinal ion profiles to obtain patterns of spatial distribution of ions and ion deposition rates (13, 28). From the spatial distribution of K^+ and Na^+ in the apical 10 mm of cotton roots, the selectivity coefficient $S_{K,Na}$ was calculated according to Pitman (26) for high Na^+ and both normal and high Ca^{2+} supply. Figure 2 indicates that the selectivity coefficient is low at high Na^+ and normal (1 mM) Ca^{2+} supply, but 10 mM Ca^{2+} increased the K^+/Na^+-selectivity dramatically in the apical 1.5 mm and less in the elongating region of the root, where the developing vacuoles probably accumulate Na^+ preferentially relative to K^+.

CALCIUM STATUS OF DEVELOPING LEAVES

High Na^+ concentrations induced Ca deficiency in corn plants (24). The symptoms disappeared after increased Ca^{2+} supply and were correlated with low Ca concentrations in the developing leaves. As the Na^+/Ca^{2+} ratio in the solution was lowered, Ca in the leaves increased while Na^+ dropped to low levels (24). In a series of papers on the effects of salt stress on Ca nutrition in barley, it was found that the Ca concentration in the shoot was decreased by soil salinity under field conditions, particularly in the younger leaves and more in the relatively salt-sensitive cv Arivat than in cv Briggs (20). In solution culture, low concentrations of NaCl inhibited Ca^{2+} transport from root to

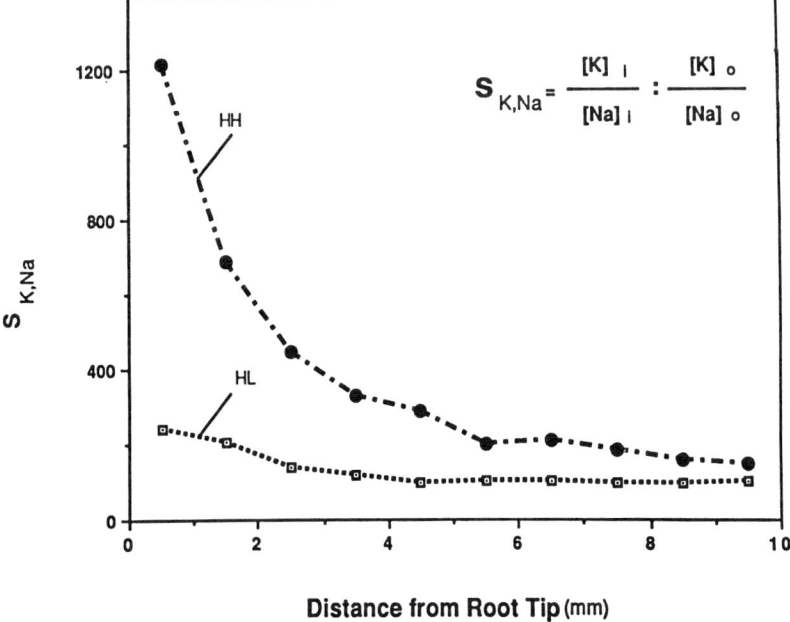

Figure 2. Longitudinal profile of the selectivity coefficient $S_{K,Na}$ in the apical 10 mm of the root of intact cotton seedlings. HL: 150 mM Na^+, 1 mM Ca^{2+}; HH: 150 mM Na^+, 10 mM Ca^{2+}. Reproduced from (16) with persmission of International Potash Institute.

shoot in cv Arivat (20). In a subsequent study using the same barley cultivars, these conclusions were essentially confirmed, and it was suggested that reduced Ca availability, coupled with high Na^+/Ca^{2+} ratios in salinized expanding leaf tissue, may contribute to shoot growth reduction in salt-stressed barley seedlings (23). Indirect support for this proposal was given by the observation of Termaat and Munns (29) that concentrated macronutrient solutions containing high Ca^{2+} concentrations were significantly less toxic than NaCl solutions of equivalent osmotic potential. Moreover, leaf elongation in salt-stressed barley seedlings was found to be improved when the Ca^{2+} supply was increased from 0.5 to 3 mM (30). In a subsequent study using the relatively salt-tolerant barley cv CM 72, these earlier conclusions could not be entirely confirmed (6). It appeared that the shoot Ca concentration or Ca transport to the shoot was not determinant for shoot growth at salt stress, unless the Ca relations of the leaf elongation zone were markedly different from the shoot as a whole (6). The latter point may be crucial in the assessment of the role of the shoot Ca status in leaf growth under salt stress, as leaf expansion is the growth process in higher plants that is most severely affected by salinity (25). In a new study on lettuce using quantitative x-ray microanalysis to assess the nutritional status of the apical shoot meristem as affected by NaCl salinization, it was found that Na$^+$ concentrations increased in tissues as close as 100 μm to the apical meristem, but Ca concentrations were reduced in the regions more basal from the apical meristem and in very young, rapidly expanding leaves (1-1.5 mm) (Lazof and Läuchli, in preparation). Future research may address the possibility that the apical shoot meristem plays a role in controlling leaf expansion. Not withstanding this possible control mechanism, the Ca status of the growing zone in leaves appears to be sensitive to salt stress; the significance of this effect in the overall plant response to high Na$^+$ requires further investigation.

DISPLACEMENT OF CALCIUM BY SODIUM FROM MEMBRANES AND INTRACELLULAR POOLS

LaHaye and Epstein (14) first proposed a mechanism by which Ca^{2+} protects membranes from the effects of high Na$^+$. They suggested that high Na$^+$ concentrations displace Ca^{2+} from the cell surface and supplemental Ca^{2+} ameliorates this effect. Recent results from my laboratory support this proposal. Cramer et al. (8), using chlorotetracycline (CTC) as a probe for membrane-associated Ca^{2+}, found increasing concentrations of Na$^+$ in the medium to decrease Ca^{2+}-CTC fluorescence in intact cotton root hairs; this effect was less pronounced when the Ca^{2+} supply was increased from 0.4 to 10 mM (Fig. 3). This was interpreted to mean that Ca^{2+} was displaced by Na$^+$ from membrane-binding sites, primarily from the plasma membrane. A

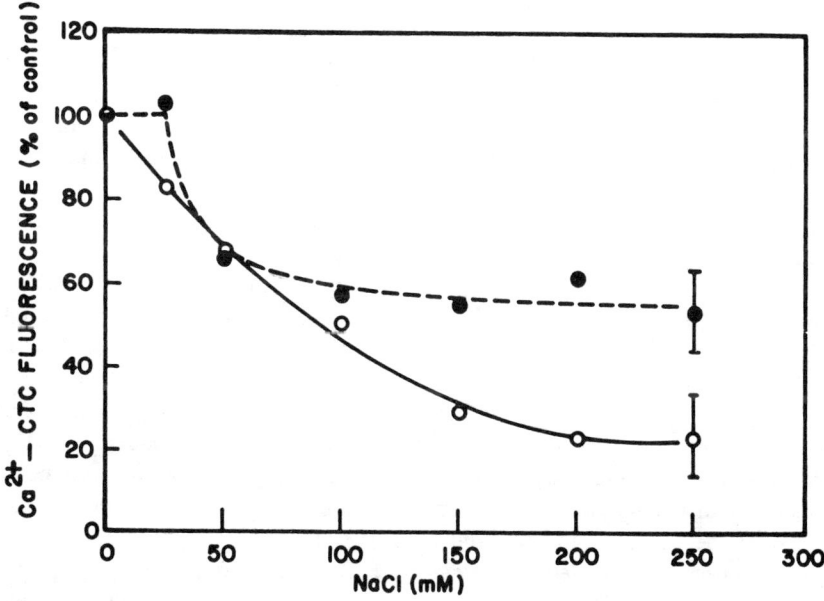

Figure 3. Effect of Na$^+$ on Ca^{2+}-CTC fluorescence in intact cotton root hairs. ○, 0.4; ●, 10 mM Ca^{2+}. Reproduced from (8) with permission of American Society of Plant Physiologists.

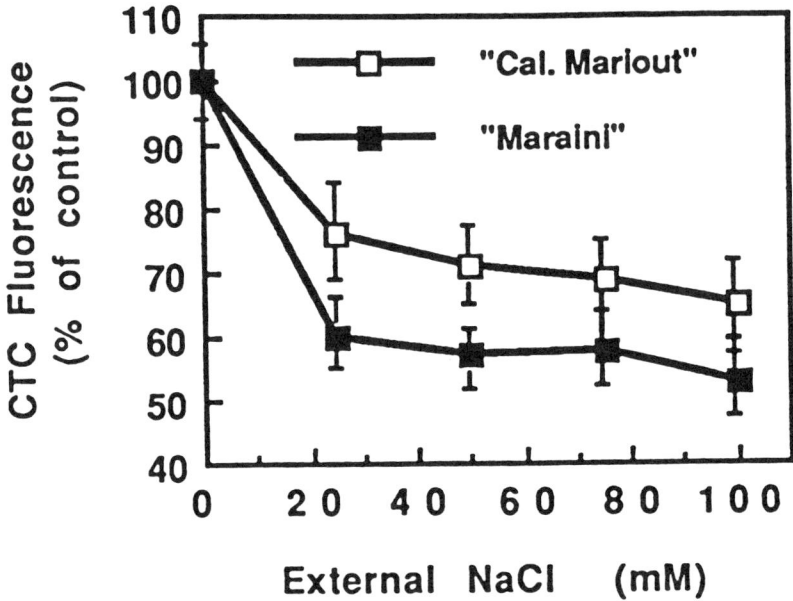

Figure 4. Effect of Na^+ on Ca^{2+}-CTC fluorescence in root protoplasts of barley cv 'Maraini' (salt sensitive) and cv 'California Mariout' (salt tolerant). Reproduced from (3) with permission of Elsevier Science Publishers/Biomedical Division.

similar response was subsequently determined in corn root protoplasts (19). Ca^{2+} interacts with membrane phospholipids that possess anionic headgroups, leading to high rigidity and surface tension of the membrane (15). Thus, Na^+-induced Ca^{2+} displacement from membranes may cause the membrane to possess lower rigidity and surface tension. Such membrane effects are likely related to altered ion fluxes. Our findings have essentially been confirmed and extended by Bittisnich et al. (3). The latter authors compared root protoplasts from two barley cultivars differing in salt tolerance (Fig. 4) and found that fluorescence quenching by NaCl was greater in the salt-sensitive cv 'Maraini' than in the tolerant cv 'California Mariout'. Thus, the high degree of salt tolerance in cv 'California Mariout' may be related to tight Ca^{2+} binding to root membranes even under salt stress.

A variety of approaches with root protoplasts later led us to believe that salt stress also displaces Ca^{2+} from intracellular membranes and pools (21). This could induce increased Ca^{2+} activity in the cytosol of the root cells. Experimental verification came from the use of the Ca^{2+} fluorescent probe Indo-1 by the technique of Bush and Jones (5). High Na^+ concentrations were shown to immediately elevate the cytosolic Ca^{2+} activity in corn root protoplasts (Fig. 5; 3). This indicates that salt stress may trigger metabolic and growth responses by way of stimulating the Ca^+ second-messenger system in the cytosol, possibly through activation of the phosphoinositide system (22).

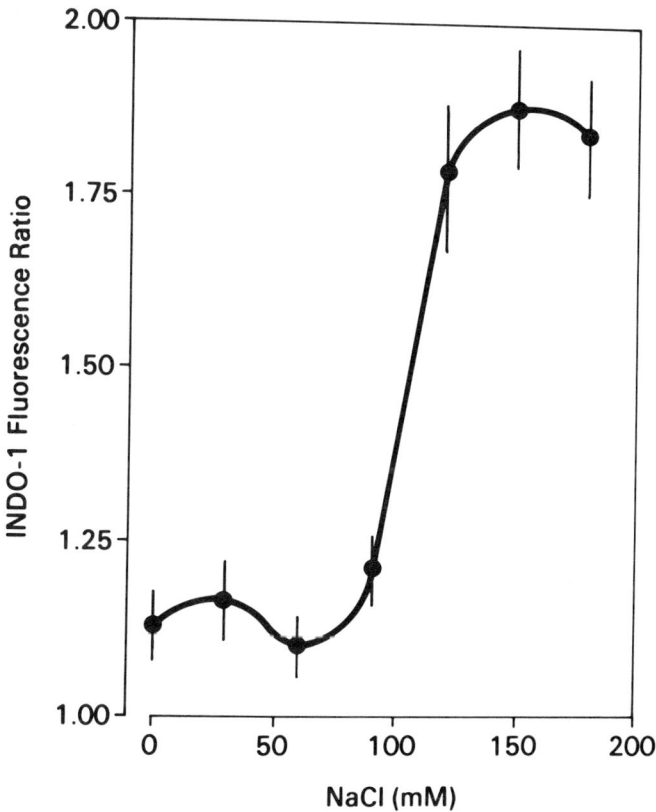

Figure 5. Effect of Na$^+$ on INDO-1 fluorescence ratios (405/480 nm) of corn root protoplasts. Reproduced from (22) with permission of American Society of Plant Physiologists.

However, these results obtained on protoplasts must be extended to walled plant cells. The use of Ca^{2+}-selective microelectrodes (11) may be very useful in this endeavor.

Displacement of membrane-associated Ca^{2+} can impair membrane functions. Cramer *et al.* (8) detected high rates of K$^+$ loss from root cells at high Na$^+$ concentrations which were partly mitigated by increased Ca^{2+} supply. Release of H$^+$ from corn roots was stimulated by moderate salinity (27), but only at relatively low (0.4 mM), and not at higher, Ca^{2+} concentration (4 mM). Root growth inhibition by salt stress may be related to the loss of nutrient ions, such as K$^+$, from the cells, but growth inhibition does not seem to be caused by reduction of cell wall acidification (31).

The disturbance of membrane function by high Na$^+$ can be clearly demonstrated by the response of the membrane potential in root cells (Fig. 6). At the lower Ca^{2+} concentration, mild salt stress depolarized the membrane

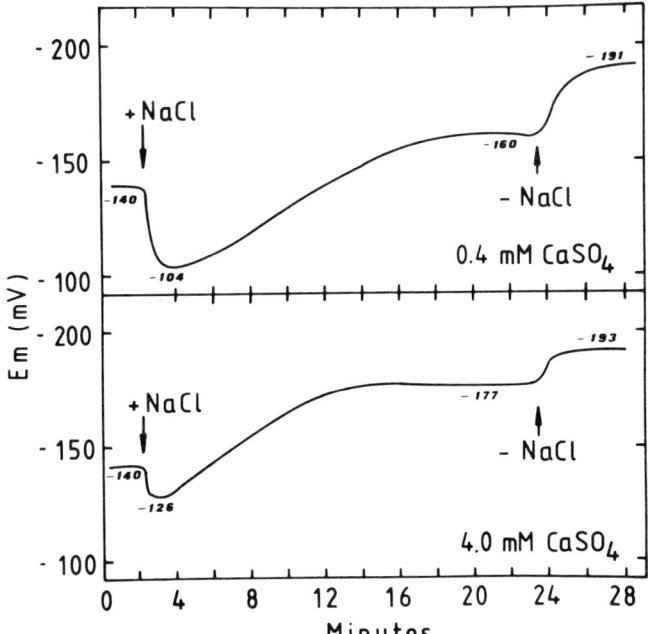

Figure 6. Response of the membrane potential Em in root cells of intact corn seedlings to Na$^+$ at two Ca^{2+} levels in the medium. Arrows indicate addition of 25 mM NaCl and exchange by NaCl-free medium. Reproduced from (18) with permission of Springer-Verlag.

potential within 1 to 2 min, followed by slower recovery. The depolarization is presumably related to Na$^+$ influx, while recovery may be ascribed to increased H$^+$ efflux due to stimulation of H$^+$-ATPase activity. As might be expected, the membrane potential was depolarized less by Na$^+$ at a 10 times greater Ca^{2+} supply. The membrane potential may be more sensitive to higher Na$^+$ concentrations and its depolarization may not be reversible under these conditions. Increased K$^+$ loss from cells is one consequence of membrane depolarization.

CELLULAR MODEL FOR ROLE OF CALCIUM IN REGULATION OF PLANT RESPONSES TO SALT STRESS

The model (Fig. 7) is mainly based on results obtained on protoplasts isolated from roots and depicts the processes that are thought to be elicited by salt stress (high Na$^+$ concentrations). Ca^{2+} is displaced by Na$^+$ at the outer surface of the plasma membrane. This event may be sensed by a receptor complex in the membrane which activates the signal transduction chain consisting of the phosphoinositide system and cytosolic Ca^{2+} activity. The latter is elevated by release of Ca^{2+} from intracellular pools, probably primarily ER. The elevated Ca^{2+} in the cytosol may trigger changes in gene expression and altered metabolism, growth, and development (see ref. 4 for second messengers in plants). High external Na$^+$ concentrations cause additional effects on the plasma membrane which are related to both Ca^{2+} displacement from the membrane and depolarization of the membrane potential. Consequences which were detected are: (*i*) inhibition of Ca^{2+} influx and stimulation of Ca^{2+} efflux, (*ii*) increased Na^+ influx, and (*iii*) enhanced K^+ leakage from the cells. The elevated Ca^{2+} in the cytosol can be lowered to the resting level by Ca^{2+} pumps at the plasma membrane, ER, and tonoplast. If the degree of salt stress is severe, the normal Ca^{2+} activity in the cytosol may not be restored, and increased Na^+ activity in the cytosol may also interfere with the regulation of metabolic processes in the cytoplasm and organelles. The protective role of supplemental Ca^{2+} against salt stress primarily operates at the plasma membrane. This cellular model must be further tested for root cells of intact plants, and also for cells of the shoot apical meristem and developing leaves.

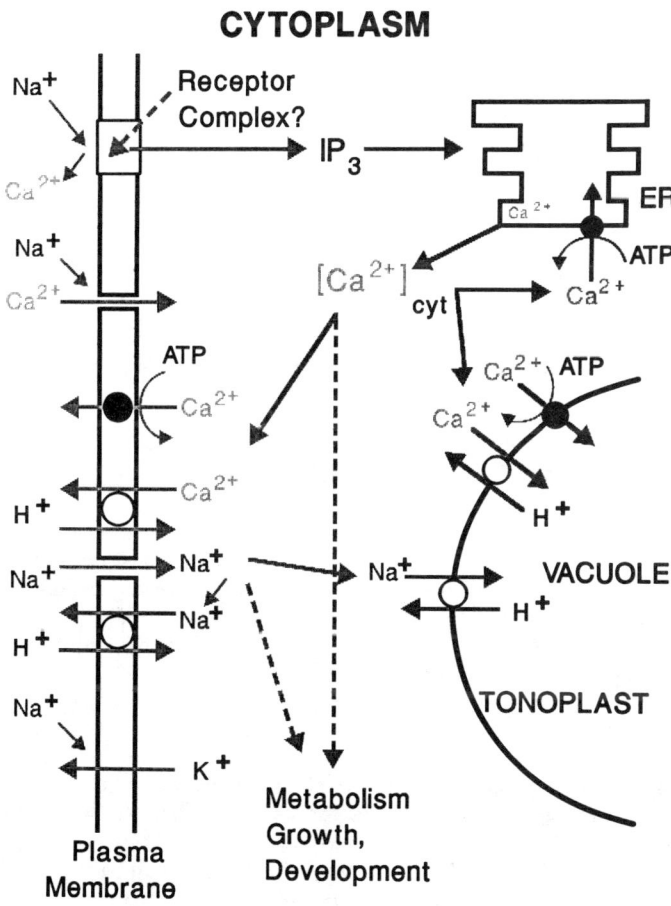

Figure 7. Cellular model for the role of Ca^{2+} in the regulation of plant responses to salt stress (high Na^+ concentrations).

ACKNOWLEDGMENTS

This work has been the product of a team effort involving my colleagues E. Epstein and V. Polito, postdoctoral researchers D. Lazof and S. Schubert, and graduate students G. Cramer, J. Lynch, and H. Zhong. I am indebted to all of them.

LITERATURE CITED

1. **Ben-Hayyim G, Kafkafi U, Ganmore-Neumann R** (1987) Role of internal potassium in maintaining growth of cultured citrus cells on increasing NaCl and $CaCl_2$ concentrations. Plant Physiol **85**: 434-439
2. **Ben-Hayyim G, Kochba J** (1982) Growth characteristics and stability of tolerance of citrus callus subjected to NaCl stress. Plant Sci Lett **27**: 87-94
3. **Bittisnich D, Robinson D, Whitecross M** (1989) Membrane-associated and intracellular free calcium levels in root cells under NaCl stress. In J Dainty, MI DeMichelis, E Marré, F Rasi-Caldogno, eds, Plant Membrane Transport: The Current Position. Elsevier, Amsterdam-New York, pp 681-682
4. **Boss WF, Morré DJ,** eds (1989) Second Messengers in Plant Growth and Development. Alan R. Liss, New York
5. **Bush DS, Jones RL** (1987) Measurement of cytoplasmic calcium in aleurone protoplasts using Indo-1 and fura-2. Cell Calcium **8**: 455-472
6. **Cramer GR, Epstein E, Läuchli A** (1989) Na-Ca interactions in barley seedlings: relationship to ion transport and growth. Plant, Cell Environ **12**: 551-558
7. **Cramer GR, Läuchli A, Epstein E** (1986) Effects of NaCl and $CaCl_2$ on ion activities in complex nutrient solutions and root growth of cotton. Plant Physiol **81**: 792-797
8. **Cramer GR, Läuchli A, Polito VS** (1985) Displacement of Ca^{2+} by Na^+ from the plasmalemma of root cells. Plant Physiol **79**: 207-211
9. **Cramer GR, Lynch J, Läuchli A, Epstein E** (1987) Influx of Na^+, K^+, and Ca^{2+} into roots of salt-stressed cotton seedlings. Plant Physiol **83**: 510-516
10. **Epstein E** (1961) The essential role of calcium in selective cation transport by plant cells. Plant Physiol **36**: 437-444
11. **Felle H** (1989) Ca^{2+}-selective microelectrodes and their application to plant cells and tissues. Plant Physiol **91**: 1239-1242
12. **Greenway H, Munns R** (1980) Mechanisms of salt tolerance in nonhalophytes. Annu Rev Plant Physiol **31**: 149-190
13. **Jeschke WD, Stelter W** (1976) Measurement of longitudinal ion profiles in single root of *Hordeum* and *Atriplex* by use of flameless atomic absorption spectroscopy. Planta **128**: 107-112
14. **LaHaye PA, Epstein E** (1969) Salt toleration by plants: enhancement with calcium. Science **166**: 395-396
15. **Landau EM, Leshem YY** (1988) Biophysical interactions of membrane anionic phospholipids with pH, calcium and auxins. J Exp Bot **39**: 1689-1697
16. **Läuchli A** (1989) Selectivity and energy-coupling of cation uptake. In Proc 21st Colloquium Int Potash Institute, Bern, Switzerland, pp 13-26
17. **Läuchli A, Epstein E** (1990) Plant responses to saline and sodic conditions. In KK Tanji, ed, Salinity Manual. American Society of Civil Engineers (in press)
18. **Läuchli A, Schubert S** (1989) The role of calcium in the regulation of membrane and cellular growth processes under salt stress. In JH Cherry, ed, Environmental Stress in Plants. Springer-Verlag, Berlin-Heidelberg, pp 131-138

19. **Lynch J, Cramer GR, Läuchli A** (1987) Salinity reduces membrane-associated calcium in corn root protoplasts. Plant Physiol **83**: 390-394
20. **Lynch J, Läuchli A** (1985) Salt stress disturbs the calcium nutrition of barley (*Hordeum vulgare* L.). New Phytol **99**: 345-354
21. **Lynch J, Läuchli A** (1988) Salinity affects intracellular calcium in corn root protoplasts. Plant Physiol **87**: 351-356
22. **Lynch J, Polito VS, Läuchli A** (1989) Salinity stress increases cytoplasmic Ca activity in maize root protoplasts. Plant Physiol **90**: 1271-1274
23. **Lynch J, Thiel G, Läuchli A** (1988) Effects of salinity on the extensibility and Ca availability in the expanding region of growing barley leaves. Bot Acta **101**: 355-361
24. **Maas EV, Grieve CM** (1987) Sodium-induced calcium deficiency in salt-stressed corn. Plant, Cell Environ **10**: 559-564
25. **Munns R, Termaat A** (1986) Whole plant responses to salinity. Aust J Plant Physiol **13**: 143-160
26. **Pitman MG** (1976) Ion uptake in plant roots. *In* U Lüttge, MG Pitman, eds, Encyclopedia of Plant Physiology (New Series), Vol 2B. Springer-Verlag, Berlin, pp 95-128
27. **Schubert S, Läuchli A** (1986) Na^+ exclusion, H^+ release, and growth of two different maize cultivars under NaCl salinity. J Plant Physiol **126**: 145-154
28. **Silk WK, Hsiao TC, Diedenhofen U, Matson C** (1986) Spatial distributions of potassium, solutes, and their deposition rates in the growth zone of the primary corn root. Plant Physiol **82**: 853-858
29. **Termaat A, Munns R** (1986) Use of concentrated macronutrient solutions to separate osmotic from NaCl-specific effects on plant growth. Aust J Plant Physiol **13**: 509-522
30. **Ward MR, Aslam M, Huffaker RC** (1986) Enhancement of nitrate uptake and growth of barley seedlings by calcium under saline conditions. Plant Pysiol **80**: 520-524
31. **Zidan I, Azaizeh H, Neumann PM** (1990) Does salinity reduce growth in maize root epidermal cells by inhibiting their capacity for cell wall acidification? Plant Physiol (in press)

MOLECULAR BIOLOGY OF PLANT P-TYPE ION-TRANSLOCATING ATPases

L.E. WIMMERS, N.N. EWING, D.J. MEYER, AND A.B. BENNETT

*Mann Laboratory, Department of Vegetable Crops,
University of California, Davis, CA 95616, USA*

Plant cell membranes are the site of several important functions including the transport of ions and metabolites between intra- and extra-cellular compartments, the perception of environmental and hormonal stimuli, and the initial steps in transduction of these signals. Membrane transport is dominated by the activity of primary ion-translocating ATPases in the tonoplast and plasma membrane as well as in many, if not all, intracellular membranes. In broad terms, the plant cell ion-translocating ATPases can be divided into two distinct classes. The first class, referred to as V-type, includes the tonoplast H^+-ATPase that also appears to be localized in a number of subcellular acidic compartments (see Taiz, this volume). The second class of ATPase is the P-type ion-translocating ATPase which is characterized by the formation of an acyl phosphorylated intermediate in its reaction cycle and a resulting sensitivity to inhibition by vanadate (33). In plants, this class of ATPase includes the plasma membrane H^+-ATPase as well as Ca^{2+}-ATPases localized in the plasma membrane, ER, and possibly other intracellular membranes. In other organisms, P-type ATPases are widespread, including the animal H^+/K^+-, NA^+/K^+-, and Ca^{2+}-ATPases, the yeast and fungal plasma membrane H^+-ATPases, and the bacterial plasma membrane K^+-ATPase (5, 17, 20, 36, 38-40).

Collectively, the plant cell ion-translocating ATPases (P and V-types) function directly to generate ion gradients for H^+ and Ca^{2+}. The H^+-ATPases generate the proton electrochemical potential gradient which, in turn, provides the driving force for the transport of numerous ions and metabolites by H^+-coupled secondary transport mechanisms and, thus, are indirectly responsible for the establishment of most cellular ion and metabolite gradients (Fig. 1) (24, 29, 30). In conjunction with ion channels for K^+, Cl^-, Ca^{2+}, and possibly other ions, the primary ion pumps and secondary transport carriers function to generate transient changes in the concentration of specific ions (12, 19). Such ion concentration transients are probably an integral part of several

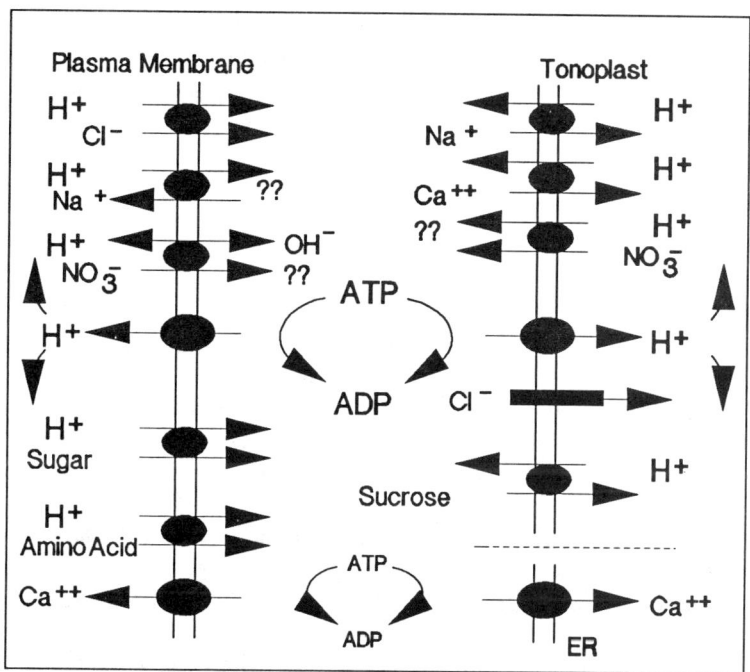

Figure 1. Schematic representation of membrane transport in higher plant cells.

signal transduction pathways. In this paper, we will review recent contributions to the understanding of the structure and function of plant cell P-type ion-translocating ATPases.

STRUCTURE AND FUNCTION OF P-TYPE ATPases

Plasma Membrane H$^+$-ATPase

The H$^+$-ATPase residing in the plant plasma membrane catalyzes the transport of H$^+$ with a stoichiometry of 1 H$^+$/ATP (30). The plasma membrane H$^+$-ATPase has a low K$_m$ for ATP (0.3-2 mM) and maintains an electrochemical gradient of approximately -150 to -250 mv and a Δ pH of approximately 2 units (33). Physiological studies suggest that the activity of the H$^+$-ATPase is kinetically regulated by auxin (14, 32), phytotoxins (25, 28), light (3, 37), internal pH (43), and probably by cytosolic Ca^{2+} concentration (33, 35).

The H$^+$-ATPase has been purified from a number of sources following differential detergent solubilization from plasma membrane-enriched microsomes (33 and references therein). The purified enzyme is comprised of a single catalytic subunit of M$_r$ 100,000, which forms a phosphorylated reaction intermediate (6). Cross-reactivity between a plant plasma membrane H$^+$-ATPase and antibodies raised to the fungal plasma membrane H$^+$-ATPase indicates weak immunological relatedness of the H$^+$-ATPases, and immunodection of the red beet plasma membrane H^+-ATPase on blots of 2-dimensional gels indicated a M_r of 100,000 and isoelectric point of 6.5 (22). Studies using radiation inactivation and glutaraldehyde-mediated cross-linking suggest that the H^+-ATPase is present in the membrane, and may function in H^+ transport, as a dimer or trimer (2, 7).

Ca^{2+}-ATPases

Plant cell Ca^{2+}-ATPases appear to be present at a much lower abundance than the H^+-ATPase (11, 30) and have been studied in less detail. The plasma membrane Ca^{2+}-ATPase maintains a steep (10^4) Ca^{2+} gradient between the cytoplasm and apoplast and has been reported to catalyze an H^+/Ca^{2+} exchange (26). The preferred substrate is MgATP, but MgGTP also drives transport at substantial rates (15). Transport is sensitive to vanadate and unaffected by protonophores (12). The K_m for MgATP has been reported to be very low (20 μM) (27). ATPase activity is Ca^{2+} dependent with a K_m for Ca^{2+} reported to be from 70 nM to 20 μM (12, 27). Calmodulin has been shown to stimulate both the affinity of the transport system for Ca^{2+} and its maximum velocity (10, 11), although other workers have observed no calmodulin stimulation (9) or have seen stimulation of Ca^{2+}-MgATP dependent ATP hydrolysis without stimulation of Ca^{2+} transport activity (27).

ATP-dependent Ca^{2+} transport is also associated with ER membranes, and the presence of a Ca^{2+}-ATPase has been inferred. Ca^{2+} uptake into ER vesicles is vanadate sensitive, protonophore insensitive, has high selectivity for MgATP with the K_m variously reported from 6 to 100 μM, and a K_m for Ca^{2+} between 70 and 370 nM (12 and references therein). The Ca^{2+}-ATPase of the ER has been reported to be insensitive to calmodulin (8).

While the plasma membrane Ca^{2+}-ATPase has not been as extensively purified or characterized as the H^+-ATPase, recent work suggests that it belongs to the P-type class of ATPases. Evans and co-workers (13) have shown an acyl-phosphorylated 140-kD polypeptide in plasma membranes from maize coleoptiles. This phosphorylation turned over rapidly in a calmodulin stimulated ATPase-dependent manner. Monoclonal antibodies raised to a partially purified preparation of plasma membrane Ca^{2+}-ATPase from maize coleoptiles inhibited calmodulin-stimulated ATPase activity and reacted with 140- and 90-kD polypeptides on protein gel blots of the purified preparation (13). Antibodies raised to the purified erythrocyte Ca^{2+}-ATPase and gold-labeled calmodulin similarly reacted with a 140-kD polypeptide on protein gel blots of the same preparation. Others have estimated the functional size of the radish plasma membrane Ca^{2+}-ATPase, by radiation inactivation analysis, to be 270 kD (27). Together these results are consistent with a Ca^{2+}-ATPase catalytic subunit of 140 kD functioning in the membrane as a dimer. Comparatively little is known about the structure of the plant ER-localized Ca^{2+}-ATPase. Recently, however, an acyl-phosphorylated 96-kD polypeptide with a rapid turnover rate was identified in red beet ER vesicles, possibly representing a reaction intermediate of the Ca^{2+}-ATPase (16).

P-TYPE ATPase GENES

A great deal has been learned about the biochemistry and function of the higher plant plasma membrane H^+- and Ca^{2+}-ATPases. Difficulties have arisen, however, because of the inherent problems in purifying membrane proteins, especially those of low abundance or high lability—problems afflicting plant P-type ATPases. As a means to: *i*) characterize the structure of P-type ATPases, *ii*) develop ATPase-specific and site-specific antibody probes, and *iii*) characterize the extent to which gene regulation may contribute to the regulation of membrane function, a productive approach has been the molecular cloning of the genes encoding P-type ATPases.

Genes of several of the P-type ATPases have now been cloned (5, 17, 20, 21, 36, 38, 40). Comparison of the deduced amino acid sequences (34, 35, 42) shows that all share 10 regions of high sequence similarity constituting over 25% of the total sequence. The degree of homology within these regions is high ($\geq$ 70%), while overall homology is generally low (25%). Exceptions include high similarity (about 85%) between isozymes of the animal Na^+/K^+- (39), plasma membrane Ca^{2+}- (38), and sarcoplasmic reticulum Ca^{2+}- ATPases (5). It is noteworthy that the animal sarcoplasmic reticulum and plasma membrane Ca^{2+}-ATPases do not share high sequence similarity. The highly conserved regions include components of the active site, a regulatory ATP-binding site, and a hydrophobic domain suggested to participate in energy transduction because of its proximity to the phosphorylation site. Hydrophobicity profiles indicate 8 to 10 transmembrane-spanning domains. Models of ATPase topology within the membrane suggest that very limited domains are exposed to the cell exterior, with large domains exposed on the cytoplasmic surface of the membrane. Not surprisingly, functional domains involved in nucleotide binding and ATP hydrolysis are located in the proposed cytosolic domains (35).

H^+-ATPase Genes

Recently, several groups have independently obtained putative cDNA and/or genomic clones of the plasma membrane H^+-ATPase in *Arabidopsis* (18, 23), tobacco (4), and tomato (Ewing *et al.*, unpublished) using similar strategies. Each clone was obtained using either oligonucleotides corresponding to portions of one or more of the highly conserved domains discussed above or a portion of the yeast H^+-ATPase gene. Of the conserved domains, the regulatory nucleotide-binding domain is the most highly conserved and is the furthest C-terminal. The presence of this sequence in the higher plant H^+-ATPase was demonstrated by sequencing of isolated peptide fragments (31). Based upon this conserved nucleotide-binding domain our laboratory designed an 8-fold degenerate oligonucleotide corresponding to this sequence.

A tomato root cDNA library comprised of 3 x 10^5 independent transformants was constructed using a vector-primer method (1) and screened with the oligonucleotide. Based on hybridization to the oligonucleotide, several cDNA clones were isolated and divided into two classes designated as LHA (*Lycopersicon* H^+-ATPase) and LCA (*Lycopersicon* Ca^{2+}-ATPase), based on cross-hybridization analysis. In addition, two distinct members of the LHA group were identified based on restriction mapping.

The two cDNAs of the LHA group (LHA1 and LHA2) were sequenced and the LHA1 cDNA found to include an in-frame initiation codon giving rise to an mRNA encoding a polypeptide possessing all of the conserved domains identified in other P-type ATPases and of the appropriate size to encode the plasma membrane H^+-ATPase. LHA2 is a partial length cDNA with an open reading frame beginning approximately 250 amino acids from the C-terminus of the deduced LHA1 sequence. LHA2 is over 90% similar to LHA1 at the nucleotide level, indicating that they are probably isozymes of the same P-type ATPase. A comparison of the similarity between LHA1, and the nucleotide sequences of the *Neurospora* H^+-ATPase (17) and a putative H^+-ATPase clone from *Arabidopsis* (20, 25) and two other P-type ATPases, is shown in Figure 2. LHA1 is 32% similar to the *Neurospora* H^+-ATPase and 19% and 16% similar to the Na^+/K^+- and a Ca^{2+}-ATPase, respectively. LHA1 shares 79% similarity with a putative plant H^+-ATPase from *Arabidopsis*.

Each of the putative plant plasma membrane H^+-ATPases share roughly the same sequence homology (35%) with the *Neurospora* H^+-ATPase gene and significantly (20-25%) less with the other P-type ATPases at the amino acid level. Each also exhibits high (70-90%) homology with the limited protein sequence obtained from tryptic peptides of the oat root plasma membrane H^+-ATPase (31), and shares very high levels of amino acid sequence homology between each other, ranging from 79 to 96%.

Support for the identification of the LHA cDNA clones was derived from *in vitro* transcription and translation of the LHA1 cDNA which gave rise to a dominant 100-kD polypeptide that is immunoprecipitable by antiserum raised to the corn root plasma membrane H^+-ATPase. In addition, the deduced amino acid sequence of LHA1 predicts a polypeptide with a molecular weight of 105 kD and an isoelectric point of 6.1. These values compare favorably with the polypeptide identified by antiserum raised to the *Neurospora* H^+-ATPase in 2-dimensional protein gel blots of red beet plasma membrane proteins (22).

It is clear that LHA1 and LHA2 encode very similar P-type ATPases. They include the highly conserved sequence domains common to P-type ATPases, and encode polypeptides of appropriate length. Further, three lines of evidence indicate that they represent the plasma membrane H^+-ATPase. First, when the nucleotide sequence of LHA1 is compared to P-type ATPases, the sequence similarity to H^+-ATPases is somewhat higher than to other P-type ATPases. Second, LHA1 and LHA2 share a high degree of sequence similarity with fragments of the oat root plasma membrane H^+-ATPase. And finally, *in vitro* translation products of the LHA1 transcript are immunoprecipitated by antiserum raised to the corn root plasma membrane H^+-ATPase. It should be noted, however, that both of these latter arguments are

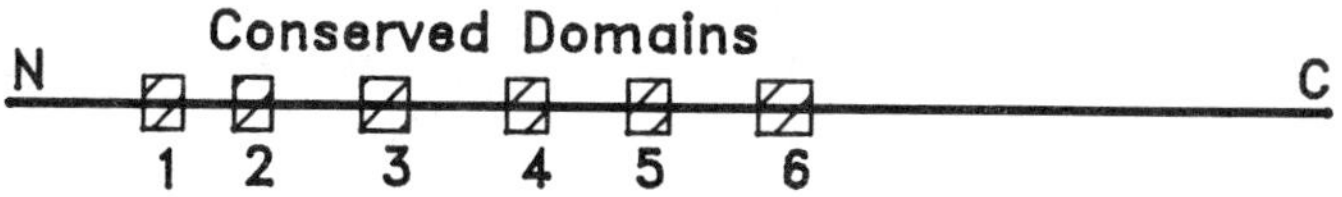

ATPase	Region:	1	2	3	4	5	6	overall
		\multicolumn{7}{c}{% SIMILARITY WITH pLHA1}
A. thalliana H$^+$-ATPase		88	97	98	62	97	91	79
N. crassa H$^+$-ATPase		56	46	49	30	36	68	32
Rabbit sarcoplasmic reticulum Ca^{2+}-ATPase		48	29	39	20	36	58	16
Sheep kidney Na$^+$/K$^+$-ATPase		46	20	29	29	33	47	19

Figure 2. Sequence similarity of LHA1 and other P-type ATPases.

dependent upon the purity of the oat and corn root H$^+$-ATPase preparations. Further proof of the identity of these clones must be obtained by their expression and functional analysis *in vitro* or in heterologous systems.

Both LHA1 and LHA2 are cDNA clones and, thus, represent products of expressed genes. In high stringency genomic Southern analysis, LHA1 and LHA2 hybridize to distinct single-restriction fragments. At low stringency, however, they hybridize to a common set of 8 to 10 genomic DNA restriction fragments, indicating the presence of 5 to 10 related genes. In Northern analysis, both LHA1 and LHA2 hybridize to 3.6 kb mRNAs present in similar abundance in roots, while the mRNA corresponding to LHA2 is relatively more abundant in leaves. The presence of multiple genes and differential abundance of two transcripts in roots and leaves suggests the potential for regulation of functionally distinct isozymes at the level of gene expression.

Ca^{2+}-ATPase Genes

A 1.7-kb member of the LCA class of cDNAs was sequenced and found to possess a region corresponding to the nucleotide-binding domain and an unusually long 3' untranslated region (860 bp). The deduced amino acid sequence of LCA was compared to other P-type ATPase sequences (Fig. 3). Comparing within the 37 amino acid nucleotide-binding domain, the sequence was between 59 and 78% identical to other ATPases, as expected for this conserved domain. Comparison of the sequences flanking the nucleotide-binding region indicated 80% similarity between LCA and a rabbit sarcoplasmic reticulum Ca^{2+}-ATPase but much lower similarity (23-41%) with other P-type ATPases. This high level of sequence similarity to the sarcoplasmic reticulum Ca^{2+}-ATPase suggests that LCA may encode a higher plant Ca^{2+}-ATPase localized in the ER or other endomembrane site.

Northern analysis of mRNA corresponding to LCA indicated the presence of three transcripts in root tissue ranging in size from 3.5 to 6.5 kb. In leaf tissue, a single, much less abundant, mRNA of 4.2 kb was detected. Southern analysis indicated a single genomic DNA restriction fragment hybridizing with LCA, suggesting that LCA may be encoded by a single gene. This result would be consistent with the production of multiple mRNAs by differential transcription of a single gene or differential processing of a single transcript.

CONCLUSIONS AND PROSPECTS

Study of plant plasma membrane P-type ATPases has been slowed by difficulties inherent in dealing with integral membrane proteins. Identification of genes encoding the plasma membrane H^+-ATPase and a Ca^{2+}-ATPase has opened the door to renewed work on existing questions, and new work on previously inaccessible questions. For example, the exhaustive identification of all genes coding for P-type ATPase isozymes within a species will make it possible to determine the relationship between transcripts and their corresponding genes, and to determine if those isozymes are functionally distinct in terms of kinetics, transport stoichiometry, or regulation. It will also be possible to assess the degree to which regulation of ATPase gene expression affects the capacity to regulate intracellular and extracellular H^+ and Ca^{2+} levels, especially in response to developmental changes and environmental stimuli.

The physiological role of the P-type ATPases can also be addressed. Antisense constructions of isozyme-specific sequences, may be used to produce

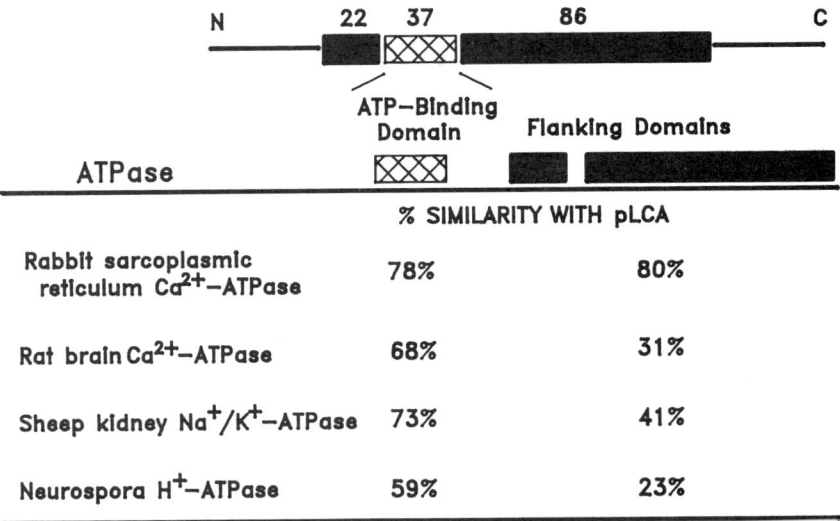

Figure 3. Sequence similarity of LCA and other P-type ATPases over a 108 amino acid domain including the nucleotide-binding region.

transgenic plants, allowing testing of hypotheses concerning the role of the H^+- and Ca^{2+}-ATPases in growth, mineral nutrition, and signal transduction. In addition, because these ATPases are the first plant plasma membrane protein genes to be cloned, it is also likely that they will stimulate work on protein targeting to the plasma membrane.

LITERATURE CITED

1. **Alexander DC** (1987) An efficient vector-primer cDNA cloning system. Methods Enzymol 154: 41-64
2. **Anton GE, Spanswick RM** (1986) Purification and properties of the H^+ translocating ATPase from the plasma membrane of tomato roots. Plant Physiol 81: 1080-1085
3. **Assmann SM, Simoncini L, Schroeder JI** (1985) Blue light activates electrogenic ion pumping in guard cell protoplasts of *Vicia faba*. Nature 318: 285-287
4. **Boutry M, Michelet B, Goffeau A** (1989) Molecular cloning of a family of plant genes encoding a protein homologous to plasma membrane H^+- translocating ATPases. Biochem Biophys Res Commun 162: 567-574.
5. **Brandl CJ, Green NM, Korczak B, MacLennan DH** (1986) Two Ca^{2+} ATPase genes: Homologies and mechanistic implications of deduced amino acid sequences. Cell 44: 597-607
6. **Briskin DP, Leonard RT** (1982) Partial characterization of a phosphorylated intermediate associated with the plasma membrane ATPase of corn roots. Proc Natl Acad Sci USA 79: 6922-6926
7. **Briskin DP, Thornley WR, Roti-Roti JL** (1985) Target molecular size of the red beet plasma membrane ATPase. Plant Physiol 78: 642-644
8. **Buckhout TJ** (1984) Characterization of Ca^{2+} transport in purified endoplasmic reticulum membrane vesicles from *Lepidium sativum* L. roots. Plant Physiol 76: 962-967
9. **Butcher RD, Evans DE** (1987) Calcium transport by pea root membranes II. Effects of calmodulin and inhibitors. Planta 172: 273-279
10. **Dieter P, Marmé D** (1980) Calmodulin activation of plant microsomal Ca^{2+} uptake. Proc Natl Acad Sci USA 77: 7311-7314
11. **Dieter P, Marmé D** (1983) The effect of calmodulin and far-red light on the kinetic properties of the mitochondrial and microsomal calcium-ion transport system from corn. Planta 159: 277-281
12. **Evans DE** (1988) Regulation of cytoplasmic free calcium by plant cell membranes. Cell Bio Int Rep 12: 383-396
13. **Evans DE, Dewey FM, Briars S-A** (1989) The calcium pumping ATPase of the plant plasma membrane. *In* J Dainty, MI DeMichelis, E Marré, F Rasi-Caldogno, eds, Plant Membrane Transport: The Current Position. Elsevier, New York, pp 231-236
14. **Gabathuler R, Cleland RF** (1985) Auxin regulation of a proton translocating ATPase in pea root plasma membrane vesicles. Plant Physiol 79: 1080-1085
15. **Giannini JL, Gildensoph LH, Reynolds-Niesman I, Briskin DP** (1987) Calcium transport in sealed vesicles from red beet (*Beta vulgaris* L.) storage tissue. I. Characterization of a Ca^{2+}-pumping ATPase associated with the endoplasmic reticulum. Plant Physiol 85: 1129-1136

16. Giannini JL, Ruiz-Cristin J, Briskin DP (1987) Calcium transport in sealed vesicles from red beet (*Beta vulgaris* L.) storage tissue. II. Characterization of $^{45}Ca^{2+}$ uptake into plasma membrane vesicles. Plant Physiol 85: 1137-1142
17. Hager KM, Mandala SM, Davenport JW, Speicher DW, Benz EJ, Slayman CW (1986) Amino acid sequence of the plasma membrane ATPase of *Neurospora crassa*: deduction from genomic and cDNA sequences. Proc Natl Acad Sci USA 83: 7693-7697
18. Harper JF, Surowy TK, Sussman MR (1989) Molecular cloning and sequence of cDNA encoding the plasma membrane proton pump (H^+-ATPase) of *Arabidopsis thaliana*. Proc Natl Acad Sci USA 86: 1234-1238
19. Hedrich R, Schroeder JI (1989) The physiology of ion channels and electrogenic pumps in higher plants. Annu Rev Plant Physiol 40: 539-569
20. Hesse JE, Wieczorek L, Altendorf K, Reicim AS, Dorus E, Epstein W (1984) Sequence homology between two membrane transport ATPases, the Kdp-ATPase of *Escherichia coli* and the Ca^{2+}-ATPase of sarcoplasmic reticulum. Proc Natl Acad Sci USA 81: 4746-4750
21. MacLennan DH, Brandle CJ, Korczak B, Green NM (1985) Amino acid sequence of a $Ca^{2+} + Mg^{2+}$-dependent ATPase from rabbit muscle sarcoplasmic reticulum, deduced from its complementary DNA sequence. Nature 316: 696-700
22. Oleski NA, Bennett AB (1987) H^+-ATPase activity from storage tissue of *Beta vulgaris*. IV. N,N'-Dicyclohexylcarbodiimide binding and inhibition of the plasma membrane H^+-ATPase. Plant Physiol 83: 569-572
23. Pardo JM, Serrano R (1989) Structure of a plasma membrane H^+-ATPase gene from the plant *Arabidopsis thaliana*. J Biol Chem 264: 8557-8562
24. Poole RJ (1978) Energy coupling for membrane transport. Annu Rev Plant Physiol 29: 437-460
25. Rasi-Caldogno F, DeMichelis MI, Pugliarello MC, Marré E (1986) H^+- pumping driven by the plasma membrane ATPase in membrane vesicles from radish: stimulation by fusicoccin. Plant Physiol 82: 121-125
26. Rasi-Caldogno F, Pugliarello MC, DeMichelis MI (1987) The Ca^{2+}- transport ATPase of plant plasma membrane catalyzes a nH^+/Ca^{2+} exchange. Plant Physiol 83: 994-1000
27. Rasi-Caldogno F, Pugliarello MC, Olivari C, DeMichelis MI (1989) Identification and characterization of the Ca^{2+}-ATPase which drives active transport of Ca^{2+} at the plasma membrane of radish seedlings. Plant Physiol 90: 1429-1434
28. Rayle DL, Cleland R (1977) Control of plant cell enlargement by hydrogen ion. Curr Top Dev Biol 11: 187-214
29. Reinhold L, Kaplan A (1984) Membrane transport of sugars and amino acids. Annu Rev Plant Physiol 35: 45-83
30. Sanders D, Slaymen CL (1989) Transport at the plasma membrane of plant cells: a review. *In* J Dainty, MI DeMichelis, E Marré, F Rasi-Caldogno, eds, Plant Membrane Transport: The Current Position. Elsevier, New York, pp 3-11
31. Schaller GE, Sussman MR (1988) Isolation and sequence of tryptic peptides from the proton-pumping ATPase of the oat plasma membrane. Plant Physiol 86: 512-516
32. Scherer GF (1984) Stimulation of ATPase activity by auxin is dependent on ATP concentration. Planta 161: 394-397
33. Serrano R (1985) Plasma Membrane ATPase of Plants and Fungi. CRC Press, Boca Raton, FL.

34. **Serrano R** (1988) Structure and function of proton translocating ATPase in plasma membranes of plants and fungi. Biochem Biophys Acta **947**: 1-28
35. **Serrano R** (1989) Structure and function of plasma membrane ATPase. Annu Rev Plant Physiol **40**: 61-94
36. **Serrano R, Kielland-Brandt MC, Fink GR** (1986) Yeast plasma membrane ATPase is essential for growth and has homology with ($Na^+ + K^+$), K^+- and Ca^{2+}-ATPases. Nature **319**: 689-693
37. **Shimazaki K, Jino M, Zeiger E** (1986) Blue light-dependent proton extrusion by guard-cell protoplasts of *Vicia faba*. Nature **319**: 324-326
38. **Shull GE, Greeb J** (1988) Molecular cloning of two isoforms of the plasma membrane Ca^{2+}-transporting ATPase from rat brain. J Biol Chem **263**: 8646-8657
39. **Shull GE, Greeb J, Lingrel JB** (1986) Molecular cloning of three distinct forms of the Na^+, K^+-ATPase α-subunits from rat brain. Biochem **25**: 8125-8132
40. **Shull GE, Lingrel JB** (1986) Molecular cloning of the rat stomach ($H^+ + K^+$)-ATPase. J Biol Chem **261**: 16788-16791
41. **Shull GE, Schwartz A, Lingrel JB** (1985) Amino acid sequence of the catalytic subunit of the ($Na^+ + K^+$) ATPase deduced from a complimentary DNA. Nature **316**: 691-695
42. **Sussman MR, Harper JF** (1989) Molecular biology of the plasma membrane of higher plants. The Plant Cell **1**: 953-960
43. **Vesper MJ, Evans ML** (1979) Nonhormonal induction of H^+ efflux from plant tissues and its correlation with growth. Proc Natl Acad Sci USA **76**: 6366-6370

CHARACTERIZATION OF THE Ca^{2+}-TRANSPORTING ATPase OF THE PLANT PLASMA MEMBRANE USING ISOLATED MEMBRANE VESICLES

DONALD P. BRISKIN, LYNNE H. GILDENSOPH, AND SWATI BASU

Department of Agronomy, University of Illinois, Urbana, IL 51801, USA

There is increasing evidence that intracellular Ca^{2+} may have an important role in the regulation of plant metabolic events and responses to the environment (11, 14). As in animal cells, it is proposed that transient increases in the level of cytoplasmic Ca^{2+} modulate the activity of key regulatory enzymes (*i.e.* protein kinases) in response to an appropriate environmental stimulus or the presence of a hormone (14 and references therein). For this signaling system to effectively function, it must be "poised" for response by the maintenance of low Ca^{2+} levels in the cytoplasm relative to the lumen of intracellular organelles or the cell exterior. This is accomplished through the action of specific transport systems responsible for mediating efflux of this cation out of the cell or its sequestration into organelles. Therefore, an overall understanding of Ca^{2+}-transport systems associated with plant cell membranes is highly relevant to understanding the basis of Ca^{2+}-mediated regulatory processes.

Transport-competent membrane vesicles isolated from plant cells have provided a convenient experimental system to investigate the biochemistry of transport systems (4, 20 for review). This approach has allowed the biochemical characterization of several Ca^{2+} transport systems associated with plant membranes, including a Ca^{2+}/H^+ antiport associated with the tonoplast (1, 19), a Ca^{2+}-translocating ATPase associated with the ER (5, 9, 12), and a Ca^{2+}-translocating ATPase associated with the plasma membrane (10, 15, 16, 21). This paper will focus on our studies of plasma-membrane-associated Ca^{2+}-translocating ATPase, using vesicles isolated from the storage tissue of red beet (*Beta vulgaris* L.). The use of this plant tissue has provided a means whereby membrane vesicles can be isolated in large quantity from the bulky storage tissue and where plasma membrane vesicles can be selectively sealed for transport studies using a procedure developed previously (8).

A Ca^{2+}-TRANSLOCATING ATPase DIRECTLY MEDIATES Ca^{2+} EFFLUX AT THE PLASMA MEMBRANE

When plasma membrane vesicles isolated from red beet storage tissue were incubated with $^{45}Ca^{2+}$, an ATP-dependent uptake of the radiolabel was observed (Fig. 1). Only a low level of radiolabel was associated with the vesicles in the absence of ATP, and the addition of the calcium ionophore A23187 rapidly discharged $^{45}Ca^{2+}$ accumulated in the presence of ATP. Since ATP cannot permeate to the interior of the vesicles (4, 20) and the active site for ATP hydrolysis would be on the side of the membrane facing the cytoplasm, these results reflect ATP-driven $^{45}Ca^{2+}$ influx in inside-out vesicles corresponding to efflux from the cell. That $^{45}Ca^{2+}$ uptake in the red beet vesicles was stimulated in the presence of gramicidin D and only slightly affected by carbonylcyanide m-chlorophenylhydrazone would suggest that transport was the result of direct coupling to ATP hydrolysis, rather than an indirect coupling to the proton electrochemical gradient established by the plasma membrane H^+-translocating ATPase (10). These two chemicals would effectively collapse the proton electrochemical gradient and inhibit transport mediated by this driving force. Further evidence that Ca^{2+} transport at the plasma membrane is mediated by a direct coupling to ATP hydrolysis is provided by our observations that $^{45}Ca^{2+}$ uptake cannot be mediated by the imposition of an artificial pH gradient (acid-interior), and that the addition of

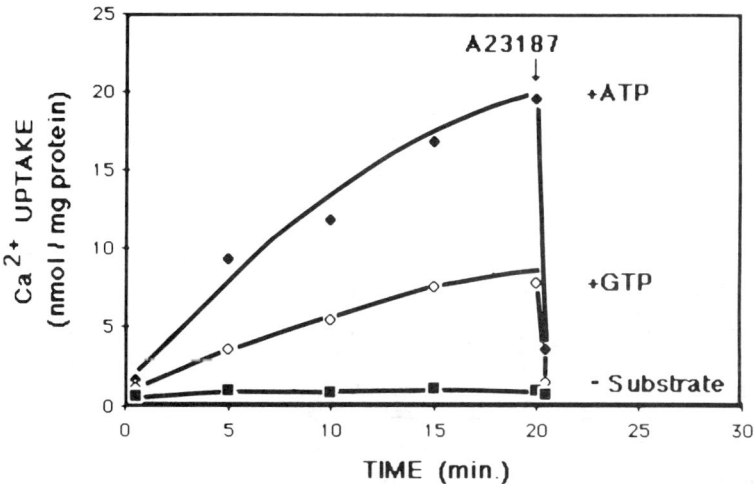

Figure 1. Uptake of $^{45}Ca^{2+}$ by red beet plasma membrane vesicles utilizing either ATP or GTP as substrate. Assays were conducted in the presence of 250 mM sorbitol, 100 mM KNO_3, 0.4 mM NaN_3, 25 mM bis-tris propane/Mes (BTP/Mes) pH 7.5, 3.75 mM $MgSO_4$, 15 µM $CaCl_2$ (2.5 µCi $^{45}Ca^{2+}$ per assay) and either 3.75 mM ATP or GTP. Both ATP and GTP were present as the BTP salt, pH 7.5. Uptake in the absence of nucleoside phosphate is also shown. At the indicated time, 0.4 µg/mL A23187 was added. Data reproduced from Williams et al. (21) with permission of American Society of Plant Physiologists.

Ca^{2+} to vesicles where a pH gradient has been established does not accelerate the collapse of the gradient (10). Taken together, these results suggest that the process of Ca^{2+} efflux at the plasma membrane is mediated by a primary Ca^{2+}-translocating ATPase, and a secondary Ca^{2+} transport system coupled to $\Delta \mu H^+$, such as a Ca^{2+}/H^+ antiport, may not be functioning.

THE PLASMA MEMBRANE Ca^{2+}-ATPase DISPLAYS CHARACTERISTICS WHICH ALLOW IT TO BE DISTINGUISHED FROM ER Ca^{2+}-ATPase

When the characteristics of Ca^{2+} transport in plasma membrane and ER vesicles from red beet are compared, it is apparent that these two transport systems display properties which allow them to be distinguished (Table I). A unique characteristic of the plasma membrane Ca^{2+}-transport system is its ability to utilize GTP as an alternative substrate for driving Ca^{2+} uptake (Fig. 1, Table I). The level of $^{45}Ca^{2+}$ uptake observed in the presence of GTP is generally about 50% of that observed in the presence of ATP, and any

Table I. *Comparison of the properties for Ca^{2+} transport in isolated vesicles from the ER and plasma membrane of red beet storage tissue.*

	Endoplasmic Reticulum[a]	Plasma Membrane[b]
Substrate Specificity	(% increase in fluorescence/min)	(nmol/min · mg prot)
ATP	4.27	3.81
GTP	0.0	2.00
CDP	0.0	0.0
CMP	0.0	0.0
Inhibitor Sensitivity	(% increase in fluorescence/min)	(nmol/min · mg prot)
Control	7.46 (100)[c]	3.24 (100)
100 μM Na_3VO_4	0.00 (0.0)	0.81 (25.0)
50 μM DCCD	2.78 (37.3)	1.72 (53.1)
30 μM DES	1.89 (25.3)	1.29 (39.8)
pH Optimum	7.25	7.5
K_m ATP	0.38 mM	0.37 mM
K_m Ca^{2+}	5.1 μM	6.0 μM

[a]Data compiled from Giannini et al. (9) where Ca^{2+} transport was determined using chlorotetracycline fluorescence.

[b]Data compiled from Giannini et al. (10) where Ca^{2+} transport was measured by the uptake of $^{45}Ca^{2+}$.

[c]Values in parenthesis indicate the percent uptake relative to the control.

accumulated radiolabel can be rapidly discharged by the addition of A23187. Uptake of $^{45}Ca^{2+}$ utilizing GTP demonstrates similar properties to that observed in the presence of ATP with respect to pH optimum, sensitivity to orthovanadate, dependence on Mg:substrate concentration and dependence on Ca^{2+} concentration (21). This ability of the plasma membrane Ca^{2+}-ATPase to utilize GTP has recently been confirmed by Rasi-Caldogno et al. (16), and these authors further demonstrate the ability of the radish seedling Ca^{2+}-ATPase to utilize ITP for driving Ca^{2+} transport. In contrast, Ca^{2+} transport mediated by the ER Ca^{2+}-ATPase shows a strong preference for utilizing ATP as the substrate for driving Ca^{2+} uptake (Table I).

Although the plasma membrane and ER Ca^{2+}-transport systems demonstrate similar pH optima and kinetic parameters for Ca^{2+} and Mg:ATP, these systems differ with respect to inhibitor sensitivity. In general, the plasma membrane Ca^{2+}-ATPase showed less sensitivity to orthovanadate, N, N'-dicyclohexylcarbodiimide (DCCD), and diethylstilbestrol (DES) than the ER-associated system. That both the plasma membrane and ER Ca^{2+}-ATPases show sensitivity to low concentrations of orthovanadate implies that these enzymes are E_1E_2-type ATPases which form phosphoenzyme intermediates during the course of their catalytic cycles (13).

CALMODULIN STIMULATION OF Ca^{2+} TRANSPORT CAN ONLY BE OBSERVED AFTER PLASMA MEMBRANE VESICLES ARE WASHED SEVERAL TIMES WITH EGTA

Stimulation of Ca^{2+} transport mediated by the plasma membrane Ca^{2+}-ATPase by calmodulin is thought to represent a characteristic feature of this transport system (7, 17). However, when $^{45}Ca^{2+}$ uptake was examined in red beet plasma membrane vesicles in the presence of calmodulin, no effect was observed with either GTP (Fig. 2A) or ATP as the substrate for driving transport (21). One possible explanation for this result could be the presence of sufficient levels of endogenous calmodulin to preclude any stimulation by the addition of exogenous calmodulin. Since repeated washing of membrane fractions with a Ca^{2+} chelator, such as EGTA, has been shown to discharge endogenous calmodulin (18), this approach was attempted with the red beet plasma membrane vesicles. As shown in Figure 2B, this explanation was confirmed since washing the vesicles four times with 4 mM EGTA allowed calmodulin stimulation of $^{45}Ca^{2+}$ uptake using either ATP or GTP as substrate.

THE ABILITY OF THE PLASMA MEMBRANE Ca^{2+}-ATPase TO UTILIZE GTP AS SUBSTRATE MAY ALLOW IDENTIFICATION OF THE CATALYTIC SUBUNIT OF THE PROTEIN

The observation that the plasma membrane Ca^{2+}-ATPase can utilize GTP to drive transport and that this activity is inhibited by low concentrations of orthovanadate (21) implies that a phosphoenzyme intermediate is involved

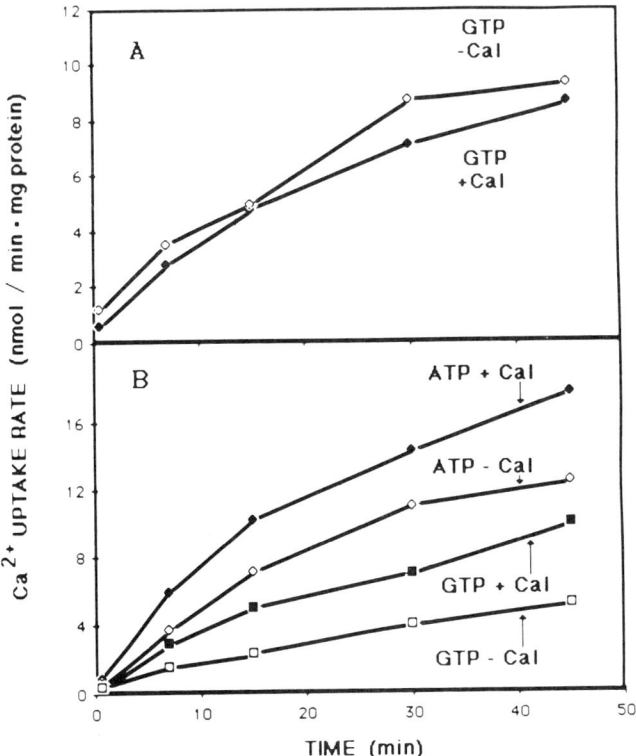

Figure 2. Effect of Calmodulin on $^{45}Ca^{2+}$ uptake in plasma membrane vesicles before and after treatment with 4 mM EGTA. Panel A: Uptake prior to EGTA washing. Panel B: Uptake after washing by suspension and centrifugation at 100,000g four times. Assays were conducted with 1 μM calmodulin as in Figure 1. Data reproduced from Williams et al. (21) with permission of American Society of Plant Physiologists.

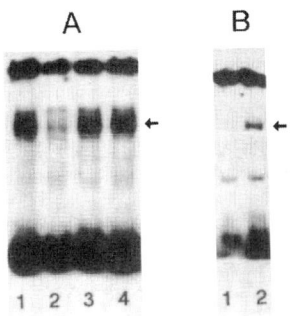

Figure 3. Phosphorylation of red beet plasma membrane vesicles using [^{32}P]-GTP. Gel A. lanes 1 & 4: 20 s phosphorylation, lane 3: 20 s phosphorylation with 0.5 mM erythrosin B, Lane 2: 20-s phosphorylation followed by 40-s chase with 100-fold excess of unlabeled GTP. Gel B. lane 1: treatment of phosphorylated protein with 0.25 N Hydroxylamine pH 5.2 lane 2: treatment with KMES pH 5.2. Data reproduced from Williams et al. (21) with permission of American Society of Plant Physiologists.

in the reaction mechanism of GTP hydrolysis and energy coupling to Ca^{2+} transport. Hence, phosphorylation of the enzyme using [α-^{32}P] GTP might allow the identification of a Ca^{2+}-ATPase phosphorylated intermediate on electrophoretic gels. Since neither the plasma membrane H^+-ATPase (3) or the ER Ca^{2+}-ATPase can utilize GTP, this approach should be specific for the plasma membrane Ca^{2+}-ATPase.

When plasma membrane vesicles were phosphorylated for 20 s using [γ-^{32}P]-GTP, a major radioactive band with a mol wt of about 100 kD was consistently observed following gel autoradiography (Fig. 3A, lanes 1 and 4). The addition of a 100-fold excess of unlabeled GTP to phosphorylation reactions at steady state (20-s) phosphorylation resulted in a rapid discharge of radioactivity associated with the 100-kD band, indicating that phosphorylation represented the turnover of an enzyme reaction intermediate (Fig. 3A, lane 2). Radioactivity associated with the 100-kD band was also discharged by treatment with 0.25 N hydroxylamine after trichloroacetic acid quenching at steady state (Fig. 3B), indicating that the protein phosphate bond represents an acylphosphate linkage characteristic of E_1E_2 type transport ATPases (13, and references therein). These results suggest that the catalytic subunit of the plasma membrane Ca^{2+}-ATPase was associated with a 100-kD peptide. However, recent studies by Briars *et al.* (2) have suggested a mol wt of 140 kD for the maize root plasma membrane Ca^{2+}-ATPase based upon attempts to identify the enzyme by immunoblots using antibodies to purified erythrocyte Ca^{2+}-ATPase. In order to confirm our identification of the red beet plasma membrane Ca^{2+}-ATPase with a 100-kD peptide, an independent means to "tag" the Ca^{2+}-ATPase catalytic peptide is needed. Our initial approach to this problem was to explore the effects of dye inhibitors of the plasma membrane Ca^{2+}-ATPase which have the potential to serve as probes for the catalytic subunit of the enzyme.

ERYTHROSIN B AND ERYTHROSIN ISOTHIOCYANATE ARE POTENT INHIBITORS OF THE PLANT PLASMA MEMBRANE Ca^{2+}-ATPase

Preliminary studies by Rasi-Caldogno *et al.* (15) have suggested that the iodinated fluorescein derivative erythrosin B might serve as a relatively specific inhibitor of the plasma membrane Ca^{2+}-ATPase when used at a concentration of less than 1 μM. The plasma membrane H^+-ATPase is inhibited by this compound, but only at much higher concentrations. Cocucci (6), found that an erythrosin B concentration of about 100 μM was required to cause 85% inhibition of the plasma membrane H^+-ATPase. When $^{45}Ca^{2+}$ uptake in red beet plasma membrane vesicles was measured over a range of erythrosin B concentrations, radiolabel uptake was inhibited 80% at an erythrosin B concentration as low as 50 nM (Fig. 4). In contrast, ATP-driven H^+ transport mediated by the plasma membrane H^+-ATPase showed very little effect with even a 20-fold higher concentration of this compound (21). Thus, erythrosin B appears to be a relatively specific inhibitor of the red beet plasma membrane Ca^{2+}- ATPase when used at low concentrations consistent with earlier studies by Rasi-Caldogno et al. (15). The observation that Ca^{2+} uptake can be completely blocked by low concentrations of erythrosin B with little effect on pH (21) would further support our proposal that Ca^{2+} efflux at the plasma membrane of red beet cells is mediated by primary transport alone and that a ΔH^+-driven secondary transport system (i.e. Ca^{2+}/H^+ antiport) might not be operative.

Although erythrosin B might act specifically with the plasma membrane Ca^{2+}-ATPase, the lack of covalent interaction would render this compound useless as a marker for the catalytic subunit. For this reason, our current studies have focused on the possible use of a chemically reactive form of erythrosin B, erythrosin isothiocyanate (EITC), as a possible means to identify the protein. This compound contains an additional isothiocyanate group which can react irreversibly with ϵ-amino groups of lysine or thiol groups of cysteine to form a covalent bond. As shown in Figure 5, EITC was also a potent inhibitor of $^{45}Ca^{2+}$ transport in red beet vesicles. For each assay, vesicles were pre-incubated with EITC for 15 min and then removed for the measurement of $^{45}Ca^{2+}$ uptake. The time course of inhibition followed a first-order relationship as shown in the inset of Figure 5. These results suggest that EITC might potentially be useful as a probe for the protein moeity of the plasma membrane Ca^{2+}-ATPase. We are currently evaluating this possibility.

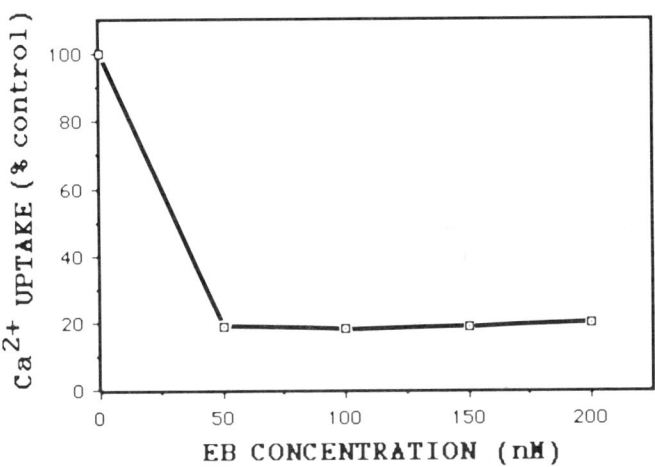

Figure 4. Effect of erythrosin B (EB) on $^{45}Ca^{2+}$ uptake in red beet plasma membrane vesicles. Assays were conducted in the presence of the indicated concentration of erythrosin B as described by Williams et al. (21).

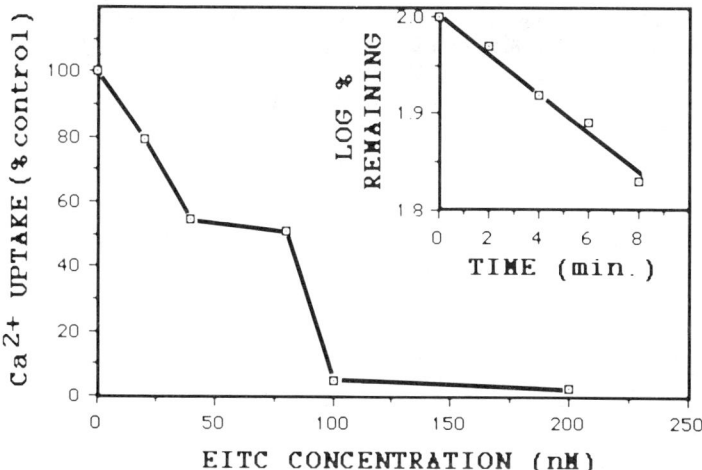

Figure 5. Effect of erythrosin isothiocyanate (EITC) on $^{45}Ca^{2+}$ uptake in red beet plasma membrane vesicles. Plasma membrane vesicles were treated with the indicated concentration of EITC for 15 min and then assayed for $^{45}Ca^{2+}$ uptake. Inset: Time course of inhibition conducted with 20 nM EITC.

SUMMARY AND PERSPECTIVE

Isolated membrane vesicles have provided a useful means for the study of Ca^{2+} transport at the plant plasma membrane. Using this approach, we have conducted preliminary biochemical characterization of the Ca^{2+}-ATPase responsible for mediating Ca^{2+}-efflux from the plant cell. Future studies will focus on understanding the structure of this important transport protein and its relationship to the mechanism of energy-coupling to Ca^{2+} transport.

LITERATURE CITED

1. **Blumwald E, Poole RJ** (1986) Kinetics of Ca^{2+}/H^+ antiport in isolated tonoplast vesicles from storage tissue of *Beta vulgaris* L. Plant Physiol **80**: 727-731
2. **Briars SA, Kessler F, Evans DE** (1988) The calmodulin-stimulated ATPase of maize coleoptiles is a 140,000 Mr polypeptide. Planta **176**: 283-285
3. **Briskin DP** (1981) Characterization of a phosphorylated intermediate associated with the plasma membrane adenosine triphosphatase of corn roots. PhD Thesis, University of California, Riverside
4. **Briskin DP** (1990) Transport in plasma membrane vesicles: approaches and perspectives. *In* I Möller, C Larsson, eds, The Plant Plasma Membrane. Structure, Function and Molecular Biology. Springer-Verlag, Berlin, pp 154-181
5. **Buckout TJ** (1984) Characterization of Ca^{2+} transport in purified endoplasmic reticulum membranes from *Lepidium sativum* L. roots. Plant Physiol **76**: 962-967
6. **Cocucci MC** (1986) Inhibition of plasma membrane and tonoplast ATPases by erythrosin B. Plant Sci **47**: 21-27
7. **Dieter P** (1984) Calmodulin and calmodulin-mediated processes in plants. Plant Cell Environ **7**: 371-380

8. **Giannini JL, Gildensoph LH, Briskin DP** (1987) Selective production of sealed plasma membrane vesicles from red beet (*Beta vulgaris* L.) storage tissue. Arch Biochem Biophys **254**: 621-630
9. **Giannini JL, Gildensoph LH, Reynolds-Niesman I, Briskin DP** (1988) Calcium transport in sealed vesicles from red beet (*Beta vulgaris* L.) storage tissue. I. Characterization of a Ca^{2+}-pumping ATPase associated with the endoplasmic reticulum. Plant Physiol **85**: 1129-1136
10. **Giannini JL, Ruiz-Cristin JL, Briskin DP** (1988) Calcium transport in sealed vesicles from red beet (*Beta vulgaris* L.) storage tissue. II. Characterization of $^{45}Ca^{2+}$ uptake into plasma membrane vesicles. Plant Physiol **85**: 1137-1142
11. **Hepler, PK, Wayne RO** (1979) Calcium and plant development. Annu Rev Plant Physiol **36**: 397-439
12. **Lew RR, Briskin DP, Wyse RE** (1986) Ca^{2+} uptake by endoplasmic reticulum from zucchini hypocotyls. The use of chlorotetracycline as a probe for Ca^{2+} uptake. Plant Physiol **82**: 47-53
13. **Pederson, PL, Carifoli E** (1987) Ion motive ATPases. I. Ubiquity, properties and significance to cell function. Trends Biochem Sci **12**: 146-150
14. **Poovaiah BW, Reddy ASN** (1987) Calcium messenger system in plants. CRC Critical Rev in Plant Sci **6**: 47-103
15. **Rasi-Caldogno F, Pugliarello MC, de Michelis MI** (1987) The Ca^{2+} transport ATPase of plant plasma membrane catalyzes a $nH+/Ca^{2+}$ exchange. Plant Physiol **83**: 994-1000
16. **Rasi-Caldogno F, Pugliarello MC, Olivari C, de Michelis MI** (1989) Identification and characterization of the Ca^{2+}-ATPase which drives active transport of Ca^{2+} at the plasma membrane of radish seedlings. Plant Physiol **90**: 1429-1434
17. **Robinson C, Larsson C, Buckout TJ** (1988) Identification of a calmodulin-stimulated (Ca^{2+} + Mg^{2+})-ATPase in a plasma membrane fraction isolated from maize (*Zea mays*) leaves. Physiol Plant **72**: 177-184
18. **Schatzman HJ** (1983) The plasma membrane calcium pump of erythrocytes and other animal cells. *In* E Carifoli, ed, Membrane Transport of Calcium. Academic Press, London, pp 41-108
19. **Schmaker KS, Sze H** (1985) A Ca^{2+}/H^+ antiport system driven by the proton electrochemical gradient of a tonoplast H^+-ATPase from oat roots. Plant Physiol **79**: 1111-1117
20. **Sze H** (1985) H^+-translocating ATPases: advances using membrane vesicles. Annu Rev Plant Physiol **36**: 175-208
21. **Williams LE, Schueler SB, Briskin DP** (1990) Further characterization of the red beet membrane Ca^{2+}-ATPase using GTP as an alternative substrate. Plant Physiol **92**: 747-754

Calcium in Plant Growth and Development, *Robert T. Leonard and Peter K. Hepler,*
Eds, 1990, The American Society of Plant Physiologists Symposium Series, Vol. 4

THE VACUOLAR ATPase:
STRUCTURE, EVOLUTION, AND PROMOTER ANALYSIS

L. TAIZ, I. STRUVE, T. RAUSCH, P. BERNASCONI, J. P. GOGARTEN,
H. KIBAK, AND S. L. TAIZ

Biology Department, University of California, Santa Cruz, CA 95064, USA
(L.T., I.S., P.B., H.K., S.L.T); Johann Wolfgang, Goethe Universitat,
Fachbereich Biologie, Botanisches Institut, Frankfurt, Germany (T.R.);
Department of Molecular and Cell Biology, University of
Connecticut, Storrs, CT 06268, USA (J.G.)

Calcium accumulates to millimolar levels in the central vacuole, whereas the concentration of cytosolic calcium is maintained in the micromolar range. Thus, the bioenergetics of calcium transport across the tonoplast is a crucial factor in the regulation of cytosolic calcium levels. The driving force for vacuolar calcium accumulation is the pH gradient generated by two electrogenic proton pumps located on the tonoplast, an ATPase and a pyrophosphatase (15). A Ca^{2+}/H^+ antiporter has been shown to mediate the exchange of calcium ions for protons (1, 5). In principle, calcium transport into the vacuole can be regulated by the activity of the primary pumps or by the activity of the antiporter. We have focused our attention on the structure, function, and regulation of the vacuolar proton-pumping ATPase (V-ATPase). To begin to understand how the enzyme might be regulated at the transcriptional level, we have analyzed the upstream region and transcripts of the gene for the 70-kD catalytic subunit. Since this enzyme is present on a variety of membranes of the secretory pathway in all eukaryotes examined thus far, it is a potentially useful marker for the evolution of the endomembrane system. Through an analysis of the sequences of the catalytic and noncatalytic subunits, insights have been gained into the prokaryotic precursor of eukaryotes.

SUBUNIT COMPOSITION AND TOPOLOGY

The V-ATPases of a number of eukaryotes, including plants, animals, and fungi, have been characterized (6, 11). In all cases, the V-ATPase has been shown to be a large, multimeric protein, about 500 kD in molecular mass, consisting of a hydrophilic complex (3, 11). The hydrophilic complex (V1) contains the catalytic site, while the integral membrane complex (Vo) functions as the proton channel. Dissociation of the V1 complex from the membrane using chaotropic anions leads to the release of two major subunits (70 and 60 kD) and several minor subunits (41, 34, and 33 kD) (3, 11). The

70-kD (A) and 60-kD (B) subunits are believed to represent the catalytic and regulatory subunits, respectively, while the minor subunits most likely attach V1 to the membrane.

The above features are similar to FoF1-type ATPases and suggest that the VoV1-ATPases may exhibit an FoF1-like "ball and stick" structure when viewed by electron microscopy. Recent electron micrographs of *Neurospora crassa* vacuolar membranes support this conclusion (3). As shown in Figure 1, negatively stained tonoplast membranes from carrot root cells are coated with F1-like particles, presumably representing V1 complexes. At higher magnification, the putative V1 complexes often appear either H-shaped or heart-shaped, with small projections emerging from the base (Fig. 1B).

EVOLUTION OF THE A AND B SUBUNITS

Sequences for the A and/or B subunits have been obtained for carrot (16), *Neurospora* (2, 4), *Arabidopsis* (9), yeast (10), and *Homo sapiens* (14). The A and B subunits have been shown to be distantly related (approximately 25% identical) to the β and α subunits, respectively, of FoF1-ATPases (12). FoF1-ATPases are present on mitochondria and chloroplast inner membranes, as well as the plasma membranes of eubacteria. The A and B subunits are also related to each other, as are the F1 α and β subunits (12). That is, the catalytic and regulatory subunits are paralogous, derived from a gene duplication event.

Remarkably, the vacuolar ATPase A and B subunits are approximately 50% identical to the α and β subunits, respectively, of archaebacterial plasma membrane ATPases (7, 8). Phylogenetic trees of the two subunits, constructed using several different algorithms, conclusively show that the archaebacterial ATPases cluster with eukaryotic vacuolar ATPases rather than with the eubacterial FoF1 ATPases (7). Because the gene duplication event that gave rise to the catalytic and noncatalytic subunits occurred prior to the separation of the three Urkingdoms from the common ancestor, the phylogenetic tree can be rooted where the two half-trees join. Assuming no lateral gene transfer, this implies that the common ancestor of all the organisms gave rise to two main branches, the archaebacteria and the eubacteria. The former subsequently diverged to give rise to eukaryotes and modern-day archaebacteria (7).

PROMOTER ANALYSIS OF THE CARROT A SUBUNIT GENE

In *Neurospora*, the genes for the A and B subunits reside on different chromosomes (2, 4), and it is likely that the situation is similar in plants. Thus, the multisubunit V-ATPase poses a number of interesting questions at the level of gene regulation. First, the genes for the various subunits are presumably expressed in a coordinate fashion to maintain the correct subunit stoichiometry, suggesting common regulatory elements in their upstream sequences. Second, the V-ATPase may represent a "housekeeping gene"

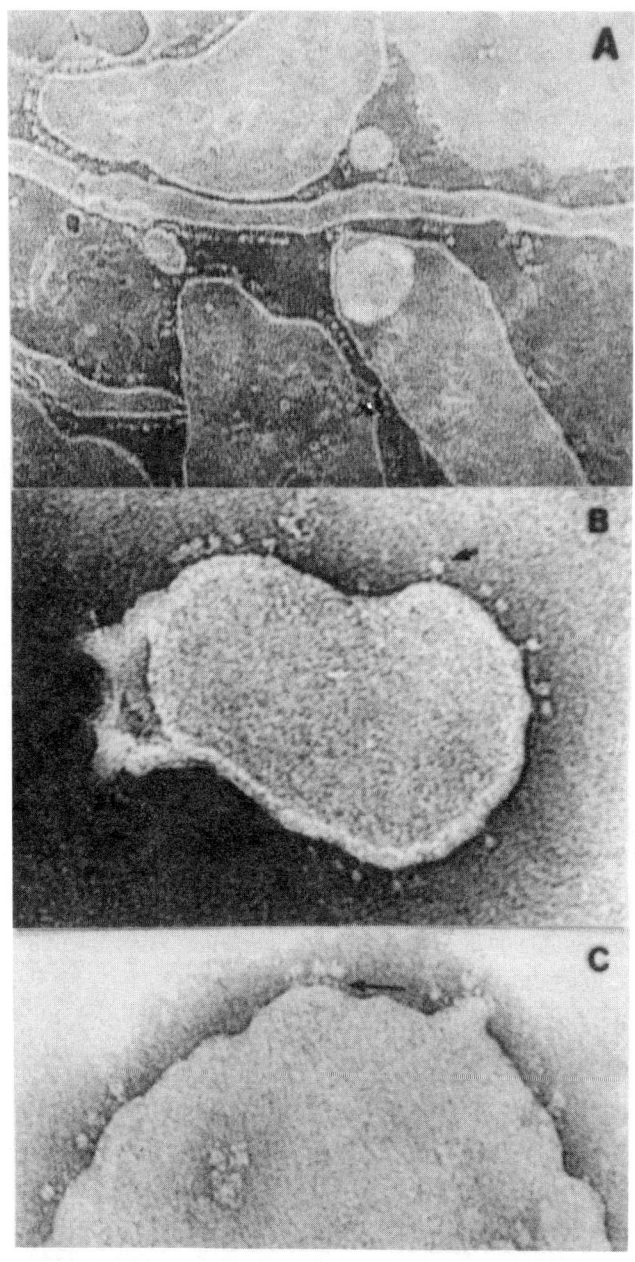

Figure 1. Carrot vacuolar ATPase visualized by negative staining with phosphotungstic acid. Vacuoles were purified from carrot root protoplasts. The arrow in B points to a putative V1 complex composed of multiple subunits. The arrow in C points to smaller structures arising from the base which may represent minor subunits.

because of its central role in solute accumulation. Little is known about the regulation of housekeeping genes in plants.

One band was detected by Southern blot analysis of carrot genomic DNA probed with a 300-bp, 5'-terminal cDNA fragment, suggesting that the A subunit gene may be present as a single copy (13). Some caution is needed, however, since it is possible that the 5' probe used in this study was unable to recognize the genes for other isoforms.

The transcription start sites of the A subunit gene were mapped by the ribonuclease protection assay (ribonuclease T1) and by primer extension analysis. Three transcriptional start sites were identified, at -224, -80, and -65, relative to the translation start site (ATG site). Sequence analysis of the 5' region demonstrated the presence of three TATA boxes positioned 25 to 31 bp upstream of each transcription start site. Two CCAAT boxes and a sequence nearly identical to the mammalian GC-box were also identified (13). In addition, several extended (TATA)n regions were located between positions -653 and -576.

The ability of the 5' region to function as a promoter was tested by fusing two different lengths of sequence to the reporter gene, β-glucuronidase (GUS). The short construct consisted of a 240-bp fragment containing all three TATA boxes and the two CCAAT boxes. The longer construct (800 bp) included the putative GC-box and the extended (TATA)n regions. The two constructs were inserted into the Ti plasmid and transferred to *Agrobacterium tumefaciens*. As a control, the GUS gene was fused to the cauliflower mosaic virus 35S promoter. Carrot root slices were then transformed, and kanamycin-resistant calli were isolated and checked for uniform expression by cytochemical assay.

Protein extracts from transformed kanamycin-resistant calli were assayed for GUS activity (Table I). Nontransformed calli showed low, but detectable, activity, whereas the cauliflower mosaic virus promoter gave very high expression, as expected. The short promoter construct gave low, but significant, expression, indicating that it contained the minimum necessary sequence to be functional as a promoter. However, the long construct gave a 6-fold higher level of expression than the short sequence. Thus, sequences farther upstream significantly increased the strength of the promoter. Further studies are needed to identify the specific elements involved, such as the putative GC-box or the (TATA)n stretches.

Table I. *Expression of β-Glucuronidase Activity by Transformed Carrot Calli*

	nmol/min/mg protein
Nontransformed Callus	0.14
Control (35S promoter)	22.27
240-bp region	0.61
800-bp region	3.73

ACKNOWLEDGMENTS

This research was funded by grants from the Department of Energy and the National Science Foundation to L.T.

LITERATURE CITED

1. **Blumwald E, RJ Poole** (1986) Kinetics of Ca/H antiport in isolated tonoplast vesicles from storage tissue of *Beta vulgaris* L. Plant Physiol **80**: 727-731
2. **Bowman BJ, R Allen, MA Wechser, EJ Bowman** (1988) Isolation of the genes encoding the *Neurospora* vacuolar ATPase: Analysis of *vma2* encoding the 57 kDa polypeptide and comparison with *vma-l*. J Biol Chem **263**: 14002-14007
3. **Bowman BJ, WJ Dschida, T Harris, EJ Bowman** (1989) The vacuolar ATPase of *Neurospora crassa* contains an F1-like structure. J Biol Chem **264**: 15606-15612
4. **Bowman EJ, K Tenney, B Bowman** (1988) Isolation of the genes encoding the *Neurospora* vacuolar ATPase: Analysis of *vma-l* encoding the 66 kDa subunit reveals homology to other ATPases. J Biol Chem **263**: 13994-14001
5. **Bush DR, H Sze** (1986) Calcium transport in tonoplast and endoplasmic reticulum vesicles isolated from cultured carrot cells. Plant Physiol **80**: 549-555
6. **Forgac M** (1989) Structure and function of the vacuolar class of ATP-driven proton pumps. Physiol Rev **69**: 765-796
7. **Gogarten JP, H Kibak, P Dittrich, L Taiz, EJ Bowman, BJ Bowman, MF Manolson, RJ Poole, T Date, T Oshima, J Konishi, K Denda, M Yoshida** (1989) The evolution of the vacuolar H^+-ATPases: Implications for the origin of eukaryotes. Proc Natl Acad Sci USA **86**: 6661-6665
8. **Gogarten JP, T Rausch, P Bernasconi, H Kibak, L Taiz** (1989) Molecular evolution of H^+-ATPases. I. *Methanococcus* and *Sulfolobus* are monophyletic with respect to eukaryotes and eubacteria. Z Naturforsch **44c**: 641-650
9. **Manolson MF, BFF Ouellette, M Filion, RJ Poole** (1988) cDNA sequence and homologies of the "57-kDa" nucleotide-binding subunit of the vacuolar ATPase from *Arabidopsis*. J Biol Chem **263**: 17987-17994
10. **Nelson H, S Mandiyan, N Nelson** (1989) A conserved gene encoding the 57-kDa subunit of the yeast vacuoles H^+ ATPase. J Biol Chem **264**: 1775-1778
11. **Nelson N** (1989) Structure, molecular genetics and evolution of vacuolar H^+ ATPases. J Bioenergetics Biomembranes **21**: 553-571
12. **Nelson N, L Taiz** (1989) The evolution of H^+-ATPases. Trends Biochem Sci **14**: 113-116
13. **Struve I, T Rausch, P Bernasconi, L Taiz** (1990) Structure and function of the promoter of the carrot V-type H^+-ATPase catalytic subunit gene. J Biol Chem (in press)
14. **Sudhof TC, VA Fried, DK Stone, PA Johnston, X-S Song** (1989) Human endomembrane H^+ pump strongly resembles the ATP-synthetase of Archaebacteria. PNAS USA **86**: 6067-6071
15. **Sze H** (1985) H^+-Translocating ATPases: advances using membrane vesicles. Annu Rev Plant Physiol **36**: 175-208
16. **Zimniac L, P Dittrich, JP Gogarten, H Kibak, L Taiz** (1988) The cDNA sequence of the 69 kDa subunit of the carrot vacuolar H^+ ATPase. Homology to the beta chain of FoF1-ATPases. J Biol Chem **263**: 9102-9112

HORMONAL REGULATION OF Ca^{2+} TRANSPORT IN MICROSOMAL VESICLES ISOLATED FROM BARLEY ALEURONE LAYERS

Douglas S. Bush and Russell L. Jones

Department of Plant Biology, University of California, Berkeley, CA 94720, USA

The plant hormone gibberellic acid (GA) induces profound changes in the metabolism and cellular structure of the barley aleurone layer (12). The changes induced by GA transform the aleurone layer from a storage tissue into a digestive gland whose primary activity is the synthesis and secretion of hydrolases, chiefly α-amylase (12). Maximal production of α-amylase and other hydrolases by the aleurone layer requires extracellular Ca^{2+} and can be blocked by Ca^{2+} removal (9) or by other agents such as ABA (12) or heat shock (1).

The mechanism by which Ca^{2+} regulates hydrolase production is poorly understood. However, Ca^{2+} levels do not alter mRNA induced by GA, as do ABA and heat shock (1, 10, 12). This suggests to us that Ca^{2+} must interact directly and reversibly with some component of the synthesis and secretion pathway. One possible site for such interaction is with a Ca^{2+} transporter. We have recently shown that enzymatic activity and structural stability of α-amylase requires micromolar levels of Ca^{2+} (8). This Ca^{2+} requirement necessitates transport of Ca^{2+} into the lumen of the ER where α-amylase is synthesized and processed prior to its secretion into the extracellular solution (15). We have, therefore, examined the effect of GA on the Ca^{2+} transport capacity of isolated membrane vesicles. Our results indicate that both GA and ABA alter the capacity of the endomembrane system to accumulate Ca^{2+} and that this capacity is correlated to α-amylase production.

MATERIALS AND METHODS

Plant Material

Isolated aleurone layers were prepared from de-embryonated barley grains (*Hordeum vulgare* L. cv Himalaya, 1985 harvest) as previously described (5). Incubation of aleurone layers was done in $CaCl_2$ (10 mM) with or without GA (5 μM) or ABA (50 μM) for 15 h, unless otherwise specified, as described by Jones and Jacobsen (14). Protoplasts were isolated from aleurone layers and resuspended in Gamborg's B5 medium containing GA (5 μM) and $CaCl_2$ (10 mM) as described by Bush et al. (5). α-Amylase activity was assayed using the starch/iodine method (16).

Membrane Isolation

Membranes were isolated from aleurone layers as described by Bush et al. (4). Briefly, aleurone layers were homogenized using a razor-blade chopper at 2°C in a buffer containing 25 mM Hepes, pH 7.4, 3 mM EDTA, 1 mM DTT, and 0.05% BSA. The resulting homogenate was filtered through two layers of Miracloth (Calbiochem, La Jolla, CA) and centrifuged at 1,000g in a Sorvall SS34 rotor (DuPont Inst., Wilmington, DE) at 2°C for 10 min. The pellet was discarded and the supernatant was centrifuged on a discontinuous sucrose gradient consisting of 50% (w/v) sucrose (1 mL), overlaid with 13% (w/v) sucrose (7 mL) in a cellulose-nitrate tube. The gradient was centrifuged at 2°C at 70,000g for 2 h in a SW 27.1 rotor (Beckman Instruments, Palo Alto, CA). Microsomal membranes were collected from the 13/50% sucrose interface with a pasteur pipette.

Calcium Transport by Microsomal Membranes

Calcium transport was assayed using the filtration method of Gross and Marm (13). Microsomal vesicles (50 μL) obtained from the discontinuous sucrose gradient were mixed with 400 μL of a reaction mixture containing 25 mM Hepes (pH 7.4), 250 mM sucrose, 10 mM potassium oxalate, 3 mM $MgSO_4$, 10 μM $CaCl_2$, 100 μM sodium azide, with or without 1 mM ATP, and $^{45}CaCl_2$ ($3.7 \cdot 10^4$ Bq mL^{-1}). Calcium transport was allowed to proceed for 20 min, after which time the samples were removed and filtered onto 0.45 μm Millipore filters under vacuum. The filters were rinsed quickly with 3.5 mL of a buffer containing 250 mM sucrose, 2.5 mM Hepes (pH 7.2), and 0.2 mM $CaCl_2$, dried, and immersed in scintillation fluid as previously described (4). Calcium transport was calculated as the difference between $^{45}Ca^{2+}$ accumulation in the presence and absence of ATP.

RESULTS

The Effect of GA and ABA on Ca^{2+} Transport and α-Amylase Activities in Isolated Membrane Vesicles

In order to examine the relationship between Ca^{2+} flux in the endomembrane system and α-amylase production, we measured Ca^{2+} transport and α-amylase activities in membrane vesicles isolated from layers treated with GA or ABA or without either hormone (Table I). The production of α-amylase by isolated barley aleurone layers is strongly influenced by GA and ABA (Table I). Incubation of aleurone layers in 5 μM GA stimulates the accumulation of α-amylase in the incubation medium by 10-fold compared to

Table I. *The Effect of GA and ABA on α-Amylase Production and on Calcium Transport in Membrane Vesicles Isolated from Barley Aleurone Layers*

Treatment[a]	Calcium Transport	α-Amylase Activity	
		Microsomal	Medium
	(pmol·min⁻¹·layer⁻¹)	(units·min⁻¹·layer⁻¹)	
GA	3.63±1.10	1.2±0.1	54±12
ABA	0.38±0.07	0.05±0.05	0.2±0.02
Ca	0.80±0.12	0.31±0.1	5.6±0.3

[a] Aleurone layers (100) were incubated in GA plus Ca^{2+} (GA), ABA plus Ca^{2+} (ABA) or Ca^{2+} alone (Ca^{2+}). After 15 h of incubation, α-amylase activity secreted into the incubation medium (MEDIUM) and in microsomal vesicles (MICROSOMAL) was measured, and calcium transport into the microsomal vesicles was measured.

incubation in Ca^{2+} alone. In contrast, incubation in 50 μM ABA inhibits the accumulation of α-amylase in the incubation medium to levels below the Ca^{2+} controls (Table I). After 15 h of incubation, the Ca^{2+} transport activity of total microsomal vesicles differed greatly between treatments, paralleling the α-amylase accumulation in the incubation media. α-Amylase and Ca^{2+} transport activities were highest in membrane vesicles isolated from GA-treated layers and lowest in those treated with ABA (Table I).

Previously we had shown that Ca^{2+} transport activity in GA-treated layers is primarily associated with vesicles derived from the ER (4). We investigated the possibility that the membranes involved in Ca^{2+} transport differed in ABA- and GA-treated layers by comparing the effect of inhibitors and ionophores on Ca^{2+} transport. Transport of Ca^{2+} in microsomes from GA-treated layers was most strongly inhibited by the ATPase inhibitor vanadate (Table II). Nitrate, an inhibitor of the H^+-ATPase of tonoplast membranes, was much less effective at inhibiting transport in these microsomes (Table II). In contrast, Ca^{2+} transport in microsomes from ABA-treated layers was equally inhibited by vanadate and nitrate (Table II). Another important difference between the effects of vanadate and nitrate in these two types of microsomes was that in microsomes from ABA-treated layers, the inhibition of nitrate and vanadate was additive: when both were present, Ca^{2+} transport was nearly abolished (Table II). In contrast, inhibition of Ca^{2+} transport in microsomes from GA-treated layers was only slightly greater in the presence of nitrate plus vanadate compared to vanadate alone (Table II). Erythrosin B, which has been reported to specifically inhibit the Ca^{2+}-ATPase of the plasma membrane (17), was slightly more effective at blocking Ca^{2+} transport in microsomes derived from ABA-treated layers, but was a relatively small fraction of the total Ca^{2+} transport in both types of microsomes (Table II). Transport of Ca^{2+} by microsomes from both GA- and ABA-treated layers was strongly inhibited by the Ca^{2+} ionophores A23187

Table II. *The Effect of Ionophores and Inhibitors on Calcium Transport in Microsomal Vesicles Isolated from Aleurone Layers Incubated in GA or ABA*

Treatment[a]	Calcium Transport Inhibition (% ± Standard Error)	
	GA	ABA
Vanadate (500 μM)	84±3	42±2
Nitrate (50 mM)	26±9	49±7
Nitrate (50 mM) plus Vanadate (500 μM)	91±7	91±11
FCCP (5 μM)	31±2	36±6
Erythrosin B (5 μM)	15±3	21±4
A23187 (1 μM)	>100	>100

[a] Aleurone layers (100) where incubated for 15 h in either GA-plus-Ca^{2+} (GA) or ABA-plus-Ca^{2+} (ABA). Microsomes were isolated and calcium transport measured without inhibitors or ionophores and with their addition (TREATMENT). Data are expressed as the difference between calcium transport in the presence or absence of inhibitor.

(Table II) and ionomycin (data not shown) indicating that in both types of vesicles, Ca^{2+} transport occurred against an electrochemical gradient.

The marked correlation between the activities of Ca^{2+} transport and α-amylase, together with the observation that α-amylase is a Ca^{2+} containing metalloenzyme, indicate that Ca^{2+} transport into the ER may be necessary for α-amylase production. In order to determine if Ca^{2+} transport could be casually related to α-amylase production, we measured the time course of increase in α-amylase and Ca^{2+} transport in microsomes derived from GA-treated layers. The stimulation of vanadate-sensitive Ca^{2+} transport that is induced by GA occurs between 0 and 6 h after treatment with GA (Fig. 1). The nitrate-sensitive component of Ca^{2+} transport is also stimulated by GA, but develops later than the vanadate-sensitive component. α-Amylase production, like vanadate-sensitive Ca^{2+} transport, increases between 0 and 6 h after treatment with GA (Fig. 1).

Since ABA decreases α-amylase production and Ca^{2+} transport, we further investigated the effect of ABA on Ca^{2+} transport by microsomes that had been pre-incubated in GA. Addition of ABA to layers that had been incubated for 10 h in GA resulted in a 40% decline in α-amylase production and a 46% decline in Ca^{2+} after 6 h.

DISCUSSION

In this paper, we have presented evidence that the plant hormones GA and ABA alter the Ca^{2+} transport capacity of the endomembrane system of barley aleurone layers. The greatest alteration of transport capacity is induced by GA, which stimulates Ca^{2+} transport 4- to 10-fold in isolated membrane

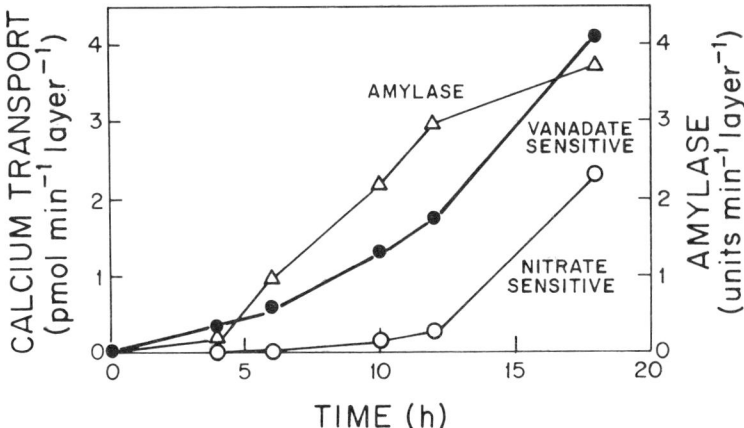

Figure 1. Time course of GA-induced increase in calcium transport and α-amylase activity in isolated microsomes. Microsomes were isolated from layers that were incubated either with or without GA for the indicated amount of time. Calcium transport and α-amylase activity in microsomes from each treatment (with or without GA). The data are expressed as the difference in the enzyme activities in the microsomes from GA-treated layers over those isolated from layers incubated without GA.

vesicles (Table I). In contrast, incubation of the aleurone in ABA (50 μM) reduces the total transport capacity of isolated membrane vesicles to a level below that of membranes isolated from aleurone layers incubated in Ca^{2+} (Table I). These changes in Ca^{2+} transport capacity are correlated both with the production of α-amylase (Fig. 1), a Ca^{2+}-containing metalloenzyme, and with ability of individual cells to regulate their cytoplasmic Ca^{2+} level *in vivo* (6, 7).

The increase in Ca^{2+} transport that is induced by incubation of the aleurone in GA appears to be primarily associated with the ER, the site of α-amylase synthesis and processing. The Ca^{2+} transport capacity of microsomes from GA-treated layers is largely vanadate sensitive (Table II), a characteristic of the Ca^{2+}-ATPase of ER membranes (2, 3). This is consistent with our previous observation that the predominant Ca^{2+} transport activity in these membranes, when resolved on a sucrose-density gradient, is located in the ER region from 30 to 32% sucrose (4). The fraction of Ca^{2+} transport inhibited by nitrate or erythrosin B, inhibitors of the tonoplast (3) and plasma membrane (17) ATPase, respectively, is much greater in microsomes isolated from ABA- or Ca^{2+}-treated layers (Table II). These data indicate that in aleurone layers not treated with GA, Ca^{2+} transport activity is roughly equal in distribution between the ER, tonoplast, and plasma membrane, whereas in GA-treated aleurone layers, Ca^{2+} is predominantly distributed into the ER.

The increase in Ca^{2+} transport into the ER that is induced by GA is consistent with the hypothesis that Ca^{2+} transport enables α-amylase production. α-Amylase is a lumenal ER protein that requires micromolar levels of Ca^{2+} for maintenance of its structure and enzymatic activity (8). Incubation of purified α-amylase in submicromolar Ca^{2+} concentrations, levels which are comparable to those in the cytosol, leads to its irreversible inactivation (8). The Ca^{2+} fluxes observed in microsomes from GA-treated layers could maintain micromolar levels of Ca^{2+} in the lumen of the ER. Indeed, it is possible to calculate the concentration of Ca^{2+} in these isolated vesicles from the observed flux, by assuming a vesicle volume of 1 μL mg^{-1} protein. This vesicle volume, which was obtained for barley root membrane vesicles (11), yields an estimated concentration of 0.5 to 1 mM Ca^{2+} in the isolated microsomal vesicles. This level is more than sufficient to maintain the structure and enzymatic activity of α-amylase. The importance of Ca^{2+} transport for α-amylase synthesis is further supported by the time course of its increase following treatment with GA (Fig 1). The increase in the vanadate-sensitive ER component of transport slightly precedes the increase in α-amylase synthesis. Thus, the development of Ca^{2+} transport could be a necessary first step in α-amylase synthesis.

In conclusion, our data show that GA greatly stimulates the Ca^{2+} transport capacity of the ER and that this increase is significant for the production of α-amylase. In contrast to GA, ABA represses the development of Ca^{2+} transport, just as it represses α-amylase production. The correlation between α-amylase and Ca^{2+} transport activities leads us to the hypothesis that, in the aleurone, hormonal regulation of Ca^{2+} transport is at the level of gene expression.

LITERATURE CITED

1. **Belanger FC, Brodl MR, Ho T-HD** (1986) Heat shock causes destabilization of specific mRNAs and destruction of endoplasmic reticulum in barley aleurone cells. Proc Natl Acad Sci USA 83: 1354-1358
2. **Buckhout TJ** (1983) ATP-dependent Ca^{2+} transport in endoplasmic reticulum isolated from roots of *Lepidium sativium* L. Planta 159: 84-90
3. **Bush DR, Sze H** (1986) Calcium transport in tonoplast and endoplasmic reticulum vesicles isolated from cultured carrot cells. Plant Physiol 80: 549-555
4. **Bush DS, Biswas AK, Jones RL** (1989) Gibberellic-acid-stimulated Ca^{2+} accumulation in endoplasmic reticulum of barley aleurone; Ca^{2+} transport and steady-state levels. Planta 178: 411-420
5. **Bush DS, Cornejo M-J, Huang C-N, Jones RL** (1986) Ca^{2+}-stimulated secretion of α-amylase during development in barley aleurone protoplasts. Plant Physiol 82: 566-574
6. **Bush DS, Jones RL** (1988) Measurement of cytoplasmic calcium in aleurone protoplasts using indo-1 and fura-2. Cell Calcium 8: 455-472
7. **Bush DS, Jones RL** (1988) Cytoplasmic calcium and α-amylase secretion from barley aleurone protoplasts. Eur J Cell Biol 46: 466-469

8. **Bush DS, Sticher L, Van Huystee R, Wagner D, Jones RL** (1989) The calcium requirement for stability and enzymatic activity of two isoforms of barley aleurone α-amylase. J Biol Chem. **264**: 19392-19398
9. **Chrispeels MJ, Varner JE** (1967) Gibberellic acid-enhanced synthesis and release of α-amylase and ribonuclease by isolated barley aleurone layers. Plant Physiol **42**: 398-406
10. **Deikman J, Jones RL** (1986) Regulation of the accumulation of mRNA of α-amylase isozymes in barley aleurone. Plant Physiol **80**: 672-675
11. **DuPont FM, Bush DS, Windle JJ, Jones RL** (1990) Calcium and proton transport in membrane vesicles from barley roots. Plant Physiol (submitted)
12. **Fincher GB** (1989) Molecular and cellular biology associated with endosperm mobilization in germinating cereal grains. Annu Rev Plant Physiol **40**: 305-346
13. **Gross J, Marme D** (1978) ATP-dependent Ca^{2+} uptake into plant membrane vesicles. Proc Natl Acad Sci USA **75**: 1232-1236
14. **Jones RL, Jacobsen JV** (1982) The role of the endoplasmic reticulum in the synthesis and transport of α-amylase in barley aleurone layers. Planta **156**: 421-432
15. **Jones RL, Robinson DG** (1989) Protein secretion in plants. New Phytol **111**: 567-597
16. **Jones RL, Varner JE** (1967) The bioassay of gibberellins. Planta **72**:155-161
17. **Rasi-Caldogno F, Pugliarello MC, Olivari C, De Michelis MI** (1989) Identification and characterization of the Ca^{2+}-ATPase which drives active transport of Ca^{2+} at the plasma membrane of radish seedlings. Plant Physiol **90**: 693-698

THE ROLE OF Ca IN MUSCLE PHYSIOLOGY

Donald M. Bers

*Division of Biomedical Sciences, University of California,
Riverside, CA 92521-0121, USA*

Ringer (30) first demonstrated that the frog's heart would not contract in the absence of extracellular Ca (Ca_o). This was indicative that Ca was involved in the regulation of contraction. However, neither the breadth of Ca involvement in muscle contraction nor the ubiquitous nature of Ca as a cellular second messenger were appreciated for many years to come. It has now become quite clear that [Ca] is the critical physiological regulator of contractile proteins and many other processes in muscle (and nonmuscle) cells. This paper focuses on Ca regulation in cardiac muscle, but also discusses the regulation of Ca and force in other muscle types, since the differences are functionally important. Ringer's experiment in frog heart actually raises the first fundamental difference. That is, that Ca entry from the extracellular space is an absolute requirement for cardiac muscle contraction, but is not required in skeletal or smooth muscle.

OVERALL Ca REGULATION IN CARDIAC MUSCLE

Figure 1a illustrates the regulation of Ca in cardiac myocytes. The wave of electrical depolarization which sweeps across the myocardial cell surface activates sarcolemmal Ca channels. Ca entry via these channels may directly activate the myofilaments to induce contraction and may also trigger the release of additional Ca from the sarcoplasmic reticulum (SR). The depolarization will also stimulate Ca entry via Na/Ca exchange (due to its electrogenic nature, 3 Na^+: 1 Ca^{2+}). This Ca may also participate in activation of myofilaments. The myofilaments are composed of an exquisitely organized matrix of proteins (myosin, actin, troponin, tropomyosin . . .) which are involved in transducing the $[Ca]_i$ change into force generation and/or muscle shortening. When myoplasmic Ca rises, Ca binds to the regulatory protein troponin on the myofilaments (in particular, to the troponin C subunit of the protein). Ca binding to troponin C induces conformational change in the other subunits of troponin (and the additional regulatory protein tropomyosin) to allow actin and myosin to interact. The framework of the sarcomere is composed of thin actin filaments. The thick filaments are composed of myosin molecules with

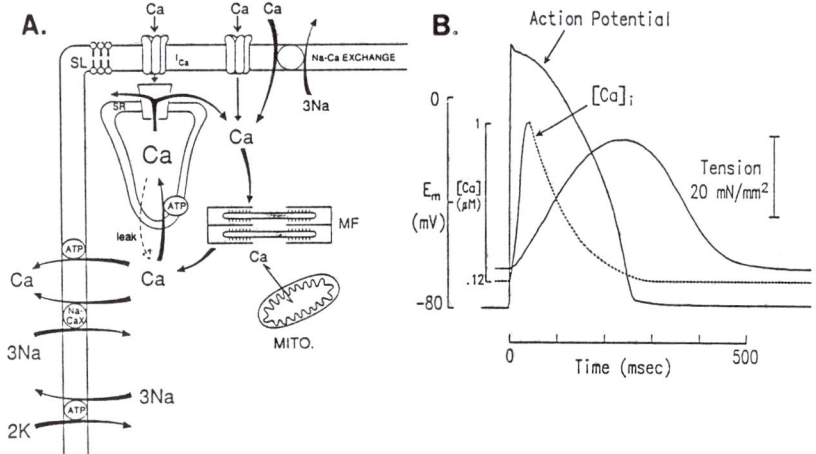

Figure 1. A, Schematic diagram of Ca fluxes in a cardiac muscle cell. B, Superimposed recordings of a cardiac action potential [Ca]$_i$ and tension in a rabbit ventricular muscle. See text for details. Panel A is reproduced from Bers (5) with permission of Kluwer Academic Publishers.

globular heads which protrude toward the thin filaments at right angles to their long axis. When myosin is allowed to bind to actin (*i.e.* due to elevated [Ca]$_i$), the globular enzymatic head uses the energy in ATP to rotate the "crossbridge" and thus generate stress, or produce actual translation (*i.e.* shortening). For relaxation to occur, Ca must be removed from the myofilaments and at least three processes contribute to this: (*i*) the SR Ca ATPase pump, (*ii*) the sarcolemmal Ca ATPase pump (which is a distinct molecular entity from the SR Ca pump), and (*iii*) the sarcolemmal Na/Ca exchange system (working in the opposite direction to that where it was responsible for Ca entry). The net Ca movements on this Na/Ca exchange system will depend both on the ionic gradients for Na and Ca, and also on the membrane potential.

Mitochondria can accumulate very large amounts of Ca (13) and under pathological conditions (*e.g.* cellular Ca overload or ischemia), large Ca phosphate precipitates can be found in mitochondria. However, the role of Ca transport by the mitochondria under physiological conditions has been more controversial. At resting levels of Ca in intact cardiac myocytes, there is very little Ca accumulation by mitochondria. Indeed, it now seems unlikely that Ca uptake and release by mitochondria are quantitatively involved in the process of excitation-contraction coupling (12). On the other hand, several important mitochondrial matrix enzymes (pyruvate dehydrogenase, α-oxoglutarate dehydrogenase, and NAD-dependent isocitrate dehydrogenase) are dependent upon [Ca] at the micromolar level (14). When either the frequency of contraction or the peak [Ca]$_i$ reached during contractions is higher, then the average cytoplasmic [Ca]$_i$ will be elevated. This, in turn, would tend to increase the average mitochondrial $[Ca]_i$ and may up-regulate mitochondrial enzymes such that the energy supply can be increased to match the demands of the higher work being performed by the heart. Thus, it seems that mitochondria play a minor role in Ca movements during contraction and relaxation *per se*. However, with slower changes in the mean cytoplasmic [Ca], mitochondrial Ca transport may play a critical role in increasing metabolic supply to meet increased metabolic demands. In more severe cellular Ca overload, mitochondria may accumulate massive amounts of Ca. However, they will do so only at the expense of ATP synthesis.

Returning to Figure 1, the (Na+K)-ATPase or Na-pump is, in large part, responsible for setting up the asymmetric ionic concentrations across the cardiac myocyte (*i.e.* low $[Na]_i$, about 10 mM *versus* $[Na]_o = 140$ mM and high intracellular [K], about 120 mM *versus* $[K]_o = 4$ mM). Under conditions where the Na-pump is inhibited, *e.g.* by the cardioactive steroid digitalis, $[Na]_i$ rises. This can have profound effects on cellular Ca homeostasis. When $[Na]_i$ rises, Ca entry via the Na/Ca exchange will tend to be more favored. While this effect is undoubtedly responsible for the increased contractile force with this class of cardiotonic drugs, it is also responsible for the cellular Ca overload and triggered arrhythmias which can result from overdose of these drugs.

In the resting cardiac myocyte, the membrane potential (E_m, inside *versus* out) is approximately -80 mV. Upon depolarization the membrane potential rapidly rises to about 30 mV, due to opening of Na- and Ca-selective channels, and then more gradually repolarizes over the time course of 200 to 300 ms. This electrical wave is known as the action potential and is responsible for triggering all of the subsequent events in the excitation-contraction coupling process. Many types of ionic channels are involved in the cardiac action potential. Shortly after the initial depolarization, there is a rise in intracellular Ca which reaches a peak much before the contraction has reached its peak.

SOURCES OF Ca FOR ACTIVATION OF CONTRACTION

The relative importance of Ca release from the SR in the activation of cardiac muscle contraction varies considerably among different cardiac preparations (3, 15). This can be most easily appreciated by comparing the effects of agents known to inhibit normal SR Ca uptake and release (caffeine and ryanodine) in different cardiac preparations (Fig. 2). It can be seen that these agents depress twitch-tension development to different extents in different cardiac preparations. This variation is apparent among different species (frog *versus* rabbit *versus* rat), at different stages of development, (neonatal *versus* adult rat), and regionally in the heart (rabbit ventricle *versus* atrium). While caffeine is well known to inhibit SR Ca transport by making the SR extremely leaky to Ca (and thereby preventing Ca uptake, 36), it has many side effects. For example, it increases myofilament Ca sensitivity (37)

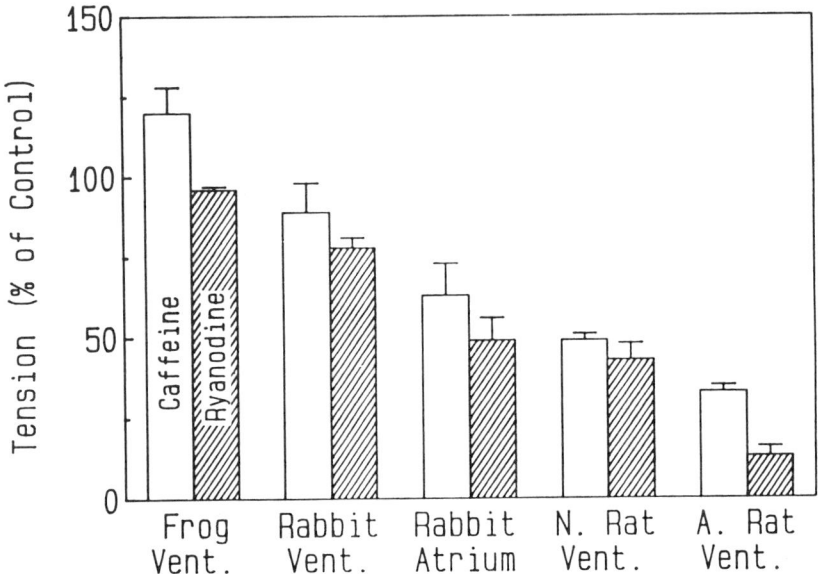

Figure 2. The effect of caffeine (10 mM) or ryanodine (100 nM) on steady-state twitch contractions (0.5 Hz at 30°C or 23°C for frog) in various cardiac muscle preparations Reproduced from Bers (5) with permission of Kluwer Academic Publishers.

and increases Ca influx (3, 34). Additionally, caffeine inhibits phosphodiesterases at the mM concentrations required for the effects on SR. Ryanodine is much more specific in its interaction with the SR (K_d = nM), but its mechanism of action seems to be somewhat more complex (7, 32). The concentration dependence of these agents does not appear to vary in different tissues, despite the difference in maximal effect. Therefore, the sensitivity of tension to depression by caffeine and ryanodine does seem, indeed, to be indicative of the relative importance of the SR in the release of Ca to activate the myofilaments. There are also ultrastructural correlates with these pharmacological dissections. For example, the most highly-developed SR network occurs in adult rat ventricle, somewhat less in rabbit atrium and ventricle, and is by far the most sparse and poorly developed in frog ventricle (5). Therefore, the virtual insensitivity of frog ventricle to ryanodine is not especially surprising. Indeed, it would appear that frog ventricular muscle can be activated entirely by Ca entering from outside the cell at each contraction (28). As a side point, it may be worth noting that most of the side effects of caffeine mentioned above would tend to increase force. Thus, in each case in Figure 2, caffeine depresses tension by a smaller amount than does ryanodine. This is particularly apparent in frog ventricle, where ryanodine has almost no effect and caffeine is slightly stimulatory. The relative importance of the SR Ca release also varies under experimental conditions in individual preparations (*i.e.* with changes in frequency, drugs, or other interventions).

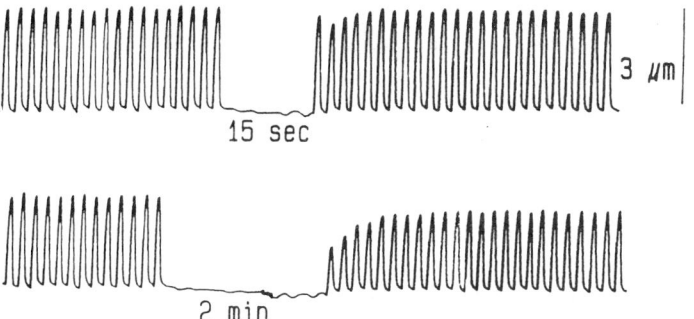

Figure 3. Post-rest recovery of twitch contractions (0.5 Hz at 23°C) in an isolated rabbit ventricular myocyte after 15 or 120 s of rest. Reproduced from Bers and Hess (9) with permission of Elsevier Scientific Publishing Co., Inc.

An example of a nonsteady-state intervention is shown in Figure 3. In the upper panel, a single rabbit ventricular myocyte stimulated at 0.5 Hz is allowed to rest for 15 s. It can be seen that there is a biphasic recovery of contractions after this brief rest period. With longer rests (2 min), the first post-rest contraction is more markedly depressed and the twitch amplitude recovers in monotonic fashion. The first post-rest contraction after a brief rest is one that is especially sensitive to ryanodine (3). Indeed, post-rest contraction after treatment with ryanodine can be very small (as opposed to relatively unaffected steady-state contractions). These types of results have indicated that the relative contribution of the SR at this first post-rest contraction may, indeed, be quite high. A corollary is that the progressively increasing contractions might have something to do with progressive increases in Ca entry. We have measured such progressive increases in Ca entry using both extracellular Ca microelectrodes to assess extracellular Ca depletion (2-4, 10, 26) and

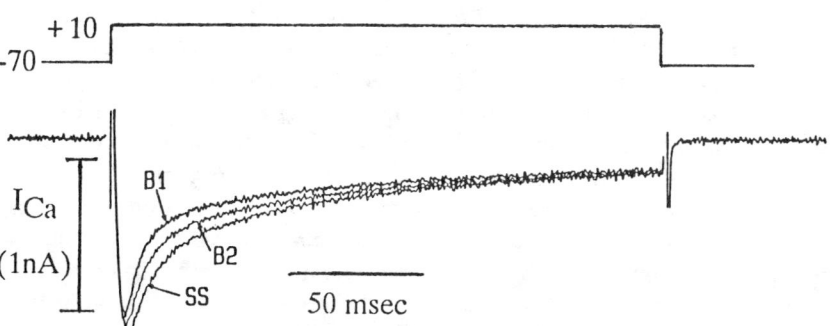

Figure 4. Ca current at steady state and after a 30-s rest measured by the whole cell variant of the patch-clamp technique in an isolated guinea-pig ventricular myocyte. The bath and intracellular solutions have Na or K replaced by tetraethylammonium and Cs to prevent contaminating ionic currents. Reproduced from Bers and Hess (9) with permission of Elsevier Scientific Publishing Co., Inc.

measurements of Ca current using the patch-clamp technique (9, 19, 25). Figure 4 shows inward Ca currents measured at a steady-state depolarizing pulse (SS) and at the first two post-rest pulses (B1 and B2) in a voltage-clamp experiment. Although the experimental conditions and solutions must be somewhat different when one is "isolating" the Ca current with such as technique, it is nonetheless clear that a "staircase" in Ca current may contribute to the "staircase" in contractions observed after a rest period. Indeed, there seems to be a delicate interplay of these two sources of Ca (Ca current and SR Ca release) in the activation of cardiac muscle contractions.

Na/Ca exchange may also be involved in Ca entry and support of contractions. Under normal conditions, the quantitative contribution of Ca entry via the Na/Ca exchange is probably very small. However, when intracellular Na is elevated, sufficient Ca may enter during the action potential via Na/Ca exchange to either activate the myofilaments directly or possibly even trigger release of Ca from the SR (8). Thus, the Na/Ca exchange may become more important in Ca influx when the transmembrane Na gradient is decreased (for example, by inhibition of the Na-pump by digitalis glycosides).

REMOVAL OF Ca FROM THE MYOPLASM DURING RELAXATION

It can be appreciated from Figure 1 that at least three processes could contribute to the reduction of Ca in the cytoplasm during relaxation (SR Ca uptake, sarcolemmal Ca ATPase pump, and Na/Ca exchange). An approach we have used to investigate the relative contributions of these three pathways is to study the rate of relaxation upon rewarming from rapid-cooling contractures (RCCs, 6). Rapid cooling of mammalian cardiac muscle from 30 to 0°C (in about 500 ms) induces a release of SR Ca and the development of a slow contracture at 0°C (the amplitude of which is indicative of the SR Ca content at the moment of cooling). During this period at 0°C, transport pumps (ATPases and Na/Ca exchange) will be largely, if not completely, inhibited and, during this period, we can modify the solution bathing the cell to preferentially effect one or more of the transport processes involved in relaxation. Figure 5 shows such an experiment. The muscle can be rewarmed in a normal Tyrode's solution (NT) and this physiological solution will reactivate all the mechanisms involved in inducing relaxation. The half-time of relaxation in this situation is about 200 ms (Fig. 5). When external Na and Ca are removed, thereby preventing Na/Ca exchange from extruding Ca, the rate of relaxation is slowed by approximately 29%. If instead we incubate with caffeine prior to rewarming, thereby inhibiting the SR Ca uptake, relaxation is slowed by about 73% on average. Finally, if both Na/Ca exchange and the SR Ca pump are inhibited (0Na + Caff), relaxation becomes extremely slow. Conclusions from this type of experiment are that Na/Ca exchange and the SR Ca pump compete for Ca during relaxation (with the SR being somewhat more effective), and further, that the sarcolemmal Ca ATPase is relatively ineffective in the process of relaxation.

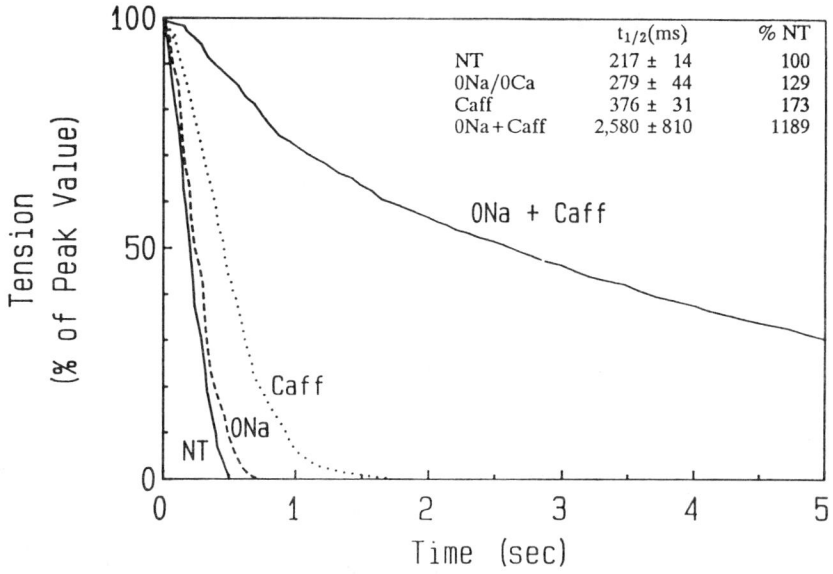

Figure 5. Relaxation upon rewarming to 30°C from an RCC in rabbit ventricular muscle, where solutions were changed from a normal Tyrode's (NT) to Na-free, Ca-free (0Na), 10 mM caffeine (Caff), or both (0Na+Caff) to suppress the Na/Ca exchange, SR Ca-pump or both. Reproduced from Bers and Bridge (6) with permission of the American Heart Association, Inc.

Figure 6 shows a different series of experiments along a similar line. In this case, we have used two sequential RCCs induced immediately after steady-state trains of twitch contractions. The key element in this experiment is that the rapid cooling releases virtually all the Ca which is available for release from the SR. Thus, during relaxation from the first RCC in NT, some Ca will be extruded from the cell (presumably by Na/Ca exchange) and some Ca will be resequestered by the SR (and be available for another RCC). The second RCC, compared to the first, will then indicate what fraction of Ca was resequestered by the SR. As can be seen in Figure 6A, the second RCC was about 25% smaller than the first. This would be consistent with Ca responsible for about 25% of the contraction being extruded by Na/Ca exchange during the first relaxation. Figure 6D shows that if the cell is exposed to Na-free, Ca-free solution during the time course of these RCCs, the second RCC is virtually identical in amplitude to the first. This suggests that indeed the decrement in the second RCC was due to Ca extrusion by Na/Ca exchange.

The SR may become gradually unloaded or emptied of Ca during prolonged rest periods, and this is illustrated by the smaller RCCs with increasing time between the RCCs in Figures 6A, B and C. That is, the amplitude of the second RCC (indicative of the SR Ca content), becomes gradually smaller with a half-time of approximately 2 min. The basis for this

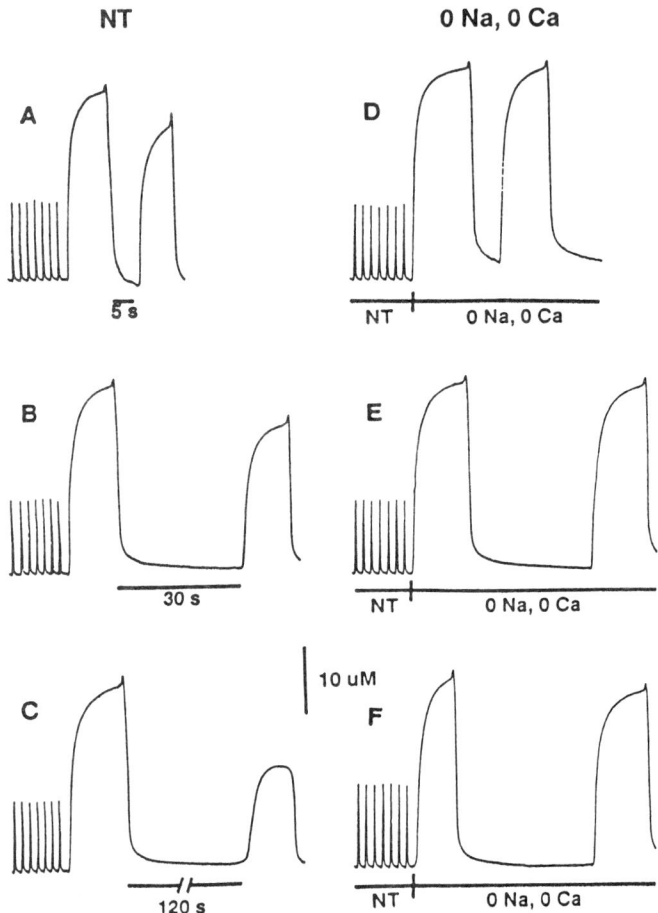

Figure 6. Paired rapid cooling contractures (RCCs, large slow contractures at approximately 0°C) in an isolated rabbit ventricular myocyte. See text for details. Reproduced from Hryshko et al. (20) with permission of the American Physiological Society.

"rest decay" of SR Ca content also seems to be intimately involved with Na/Ca exchange. This is indicated by Figure 6D-F where, in the absence of external Na, even with a 2-min rest period, there is virtually no decrement in SR Ca content. Thus, this more gradual process of rest-decay from the SR also seems to depend critically on Na/Ca exchange (and not the sarcolemma Ca-pump). Referring back to Figure 1, this pathway can be clearly understood. That is, during rest, there must be a finite leak of Ca from the SR to the myoplasm, at which point several transport processes will compete for that Ca (including re-uptake into the SR). If some fraction of Ca is extruded from the cell via Na/Ca exchange, this will serve to gradually deplete the SR of Ca during the period of rest.

MECHANISMS OF EXCITATION-CONTRACTION COUPLING IN DIFFERENT MUSCLE TYPES

Figure 7 illustrates the mechanisms of excitation-contraction (E-C) coupling in cardiac, skeletal, and smooth muscle. This comparison is especially useful because it appears, at the present time, that these three different muscle types have distinct mechanisms of E-C coupling. While I have discussed Ca entry via Ca channels in cardiac muscle as a source of activator Ca, such Ca entry via Ca channels also appears to be absolutely required for induction of Ca release from the cardiac SR (29). This model, known as Ca-induced Ca-release, has been extensively characterized over the past 15 years by Fabiato (*e.g.* 15-18). In this model, a small amount of Ca

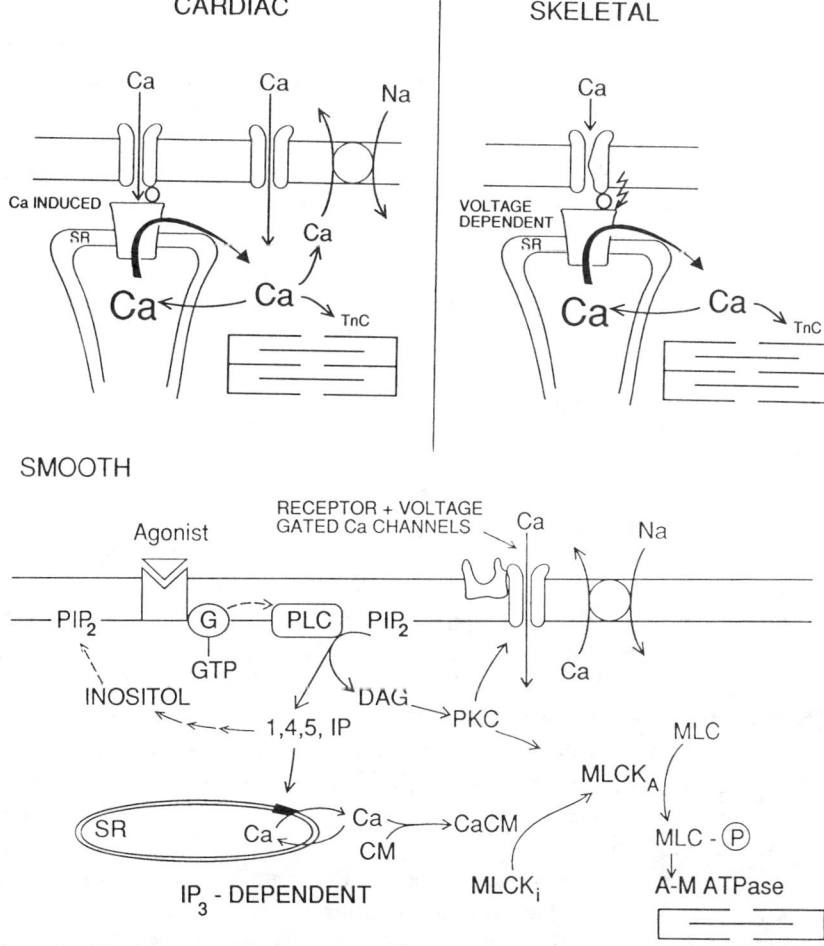

Figure 7. Mechanisms of E-C coupling in cardiac (A), skeletal (B), and smooth muscle (C). See text for details. Reproduced from Bers (5) with permission of Kluwer Academic Publishers.

entry via the Ca channel induces release of additional Ca from the SR (in a graded fashion). As $[Ca]_i$ rises, the SR Ca release channel appears to inactivate (preventing further positive feedback). The SR Ca release channel in both cardiac and skeletal muscle is also the ryanodine receptor and has recently been purified, cloned, reconstituted into bilayer membranes, and identified structurally as the "foot" protein that spans the gap between the SR and the transverse tubule membrane (an infolding of the sarcolemma, 21-24, 33). Furthermore, there is both biochemical and ultrastructural evidence that the sarcolemmal Ca channel [or dihydropyridine (DHP) receptor] may be in a closely apposed physical location overlying the SR Ca release channel (11, *e.g.* see Figs. 1 and 7). Thus, the pieces are falling into place to support this model of cardiac E-C coupling although there are still many unanswered questions about the nature of the molecular interactions and actual mechanism of release.

While there is considerable support for the above Ca-induced Ca-release hypothesis in cardiac muscle, Ca entry does not appear to be required for E-C coupling in skeletal muscle (1). In skeletal muscle, it appears that sarcolemmal membrane depolarization itself (rather than Ca current) is responsible for inducing SR Ca release (27). However, the DHP receptor still appears to be centrally involved. That is, Ca-channel antagonist DHPs can block skeletal muscle E-C coupling (and the charge movement which is thought to underlie the process, 31). A unifying hypothesis is that the DHP-receptor (and Ca-channel protein) is the key element in both muscle types. In particular, Ca-current through the channel is required to induce cardiac SR Ca release, where in skeletal muscle, the voltage-dependent change in the DHP receptor is sufficient to induce SR Ca release without requiring any Ca-current.

Although smooth muscle has sarcolemmal Ca channels (both voltage and receptor activated), Na/Ca exchange and is ryanodine-sensitive, the central mechanism of E-C coupling appears to be quite different in smooth muscle (35, Fig. 7C). A receptor agonist (*e.g.* a_1-adrenergic in the case of most vascular smooth muscle) activates a receptor and a GTP-binding protein coupled to that receptor. This activates phospholipase C, resulting in the production of inositol 1,4,5-trisphosphate (IP_3) and diacylglycerol (DAG). IP_3 is able to stimulate release of Ca from the SR, and DAG can activate protein kinase C, which can modify the contractile proteins and ion channels. The Ca-sensitive molecular "switch" allowing actin and myosin to interact is also different in smooth muscle. That is, Ca plus calmodulin activates a myosin light chain kinase to phosphorylate a myosin light chain which allows actin and myosin to interact. This contrasts with the more direct role of Ca via troponin C in striated muscle.

In conclusion, the key mechanisms in E-C coupling appear to be: *i*) Ca-induced Ca-release from the SR and Ca influx in cardiac muscle, *ii*) voltage-dependent SR Ca release in skeletal muscle, and *iii*) IP_3-induced SR Ca release (and Ca influx) in smooth muscle. This, however, is undoubtedly an oversimplification, since arguments have been made for every one of these processes in every one of the muscle types at some time.

LITERATURE CITED

1. **Armstrong CM, Benzanilla FM, Horowitz P** (1972) Twitches in the presence of ethyleneglycol-b-s-(b-amino-ethylether)-N'N' tetracetic acid. Biochim Biophys Acta **267**: 605-608
2. **Bers DM** (1983) Early transient depletion of extracellular [Ca] during individual cardiac muscle contractions. Am J Physiol **244**: H462-H468
3. **Bers DM** (1985) Ca influx and SR Ca release in cardiac muscle activation during postrest recovery. Am J Physiol. **248**: H366-H381
4. **Bers DM** (1987) Mechanisms contributing to the cardiac inotropic effect of Na-pump inhibition and reduction of extracellular Na. J Gen Physiol **90**: 479-504
5. **Bers DM** (1990) Excitation-Contraction Coupling and Cardiac Contractile Force. Kluwer Academic Press, The Netherlands (in press)
6. **Bers DM, Bridge JHB** (1989) Relaxation of rabbit ventricular muscle by Na-Ca exchange and sarcoplasmic reticulum Ca-pump: Ryanodine and voltage sensitivity. Circ Res **65**: 334-342
7. **Bers DM, Bridge JHB, MacLeod KT** (1987) The mechanism of ryanodine action in cardiac muscle assessed with Ca selective microelectrodes and rapid cooling contractures. Can J Physiol Pharmacol **65**: 610-618
8. **Bers DM, Christensen DM, Nguyen TX** (1988) Can Ca entry via Na-Ca exchange directly activate cardiac muscle contraction? J Mol Cell Cardiol **20**: 405-414
9. **Bers DM, Hess P** (1988) The influence of rest periods on calcium currents and contractions in isolated ventricular myocytes from guinea-pig and rabbit hearts. In WA Clark, RS Decker, TK Borg, eds, Biology of Isolated Adult Cardiac Myocytes. Elsevier Science Publishing Co, Inc, New York, pp 410-413
10 **Bers DM, MacLeod KT** (1986) Cumulative extracellular Ca depletions in rabbit ventricular muscle monitored with Ca selective microelectrodes. Circ Res **58**: 769-782
11. **Block BA, Imagawa T, Campbell KP, Franzini-Armstrong C** (1988) Stuctural evidence for direct interaction between the molecular components of the transverse tubule/sarcoplasmic reticulum junction in skeletal muscle. J Cell Biol **107**: 2587-2600
12. **Carafoli E** (1987) Intracellular calcium homeostasis. Annu Rev Biochem **56**: 395-433
13. **Carafoli E, Lehninger AL** (1971) A survey of the interaction of calcium ions with mitochondria from different tissues and species. Biochem J **122**: 618-690
14. **Denton RM, McCormack JG** (1985) Ca^{2+} transport by mammalian mitochondria and its role in hormone action. Am J Physiol **249**: E543-E554
15. **Fabiato A** (1982) Calcium release in skinned cardiac cells: Variations with species, tissues, and development. Fed Proc **41**: 2238-2244
16. **Fabiato A** (1985) Rapid ionic modifications during the aequorin-detected calcium transient in a skinned canine cardiac Purkinje cell. J Gen Physiol **85**: 189-246, 1985.
17. **Fabiato A** (1985) Time and calcium dependence of activation and inactivation of calcium-induced release of calcium from the sarcoplasmic reticulum of a skinned canine cardiac Purkinje cell. J Gen Physiol **85**: 247-290
18. **Fabiato A** (1985) Simulated calcium current can both cause calcium loading in and trigger calcium release from the sarcoplasmic reticulum of a skinned canine cardiac Purkinje cell. J Gen Physiol **85**: 291-320

19. Hryshko LV, Bers DM (1988) Control of Ca current staircase after rest in rabbit ventricular myocytes. Physiologist 31: A131
20. Hryshko LV, Stiffel VM, Bers DM (1989) Rapid cooling contractures as an index of SR Ca content in rabbit ventricular myocyte. Am J Physiol 257: H1369-H1377
21. Imagawa T, Smith JS, Coronado R, Campbell KP (1987) Purified ryanodine receptor from skeletal muscle sarcoplasmic reticulum is the Ca^{2+}-permeable pore of the calcium release channel. J Biol Chem 262: 16636-16643
22. Inui M, Saito A, Fleischer S (1987) Purification of the ryanodine receptor and identity with feet structures of junctional terminal cisternae of sarcoplasmic reticulum from fast skeletal muscle. J Biol Chem 262: 1740-1747
23. Inui M, Saito A, Fleischer S (1987) Isolation of the ryanodine receptor from cardiac sarcoplasmic reticulum and identity with the feet structures. J Biol Chem 262: 15637-15642
24. Lai FA, Erickson H, Block BA, Meissner G (1987) Evidence for a junctional feet-ryanodine receptor complex from sarcoplasmic reticulum. Biochem Biophys Res Commun 143: 704-709
25. Lee KS (1987) Potentiation of the calcium-channel currents of internally perfused mammalian heart cells by repetitive depolarization. Proc Natl Acad Sci USA 84: 3941-3945
26. MacLeod KT, Bers DM (1987) The effects of rest duration and ryanodine on extracellular calcium concentration in cardiac muscle from rabbits. Am J Physiol 253: C398-C407
27. Melzer W, Schneider MF, Simon B, Szucs G (1986) Intramembrane charge movement and Ca release in frog skeletal. J Physiol 373: 481-511
28. Morad M, Cleemann L (1987) Role of Ca^{2+} channel in development of tension in heart muscle. J Mol Cell Cardiol 19: 527-553
29. Näbauer M, Callewaert G, Cleemann L, Morad M (1989) Regulation of calcium release is gated by calcium turrent, not gating charge, in cardiac myocytes. Science 244: 800-803
30. Ringer SA (1883) A further contribution regarding the influence of different constituents of the blood on the contraction of the heart. J Physiol Lond 4: 29-42
31. Rios E, Brum G (1987) Involvement of dihydropyridine receptors in excitation-contraction coupling in skeletal muscle. Nature 325: 717-720
32. Sutko JL, Ito K, Kenyon JL (1985) Ryanodine: A modifier of sarcoplasmic reticulum calcium release. Biochemical and functional consequences of its actions on striated muscle. Fed Proc 44: 2984-2988
33. Takeshima H, Hishimura S, Matsumoto T, Ishida H, Kangawa K, Minamino N, Matsuo H, Ueda M, Hanaoka M, Hirose T, Numa S (1989) Primary structure and expression from complementary DNA of skeletal muscle ryanodine receptor. Nature 339: 439-445
34. Tseng G (1988) Calcium current restitution in mammalian ventricular myocytes is modulated by intracellular calcium. Circ Res 63: 468-482
35. van Breemen C, Saida K (1989) Cellular mechanisms regulating $[Ca^{2+}]_i$ smooth muscle. Annu Rev Physiol 51: 315-329
36. Weber A, Herz R (1968) The relationship between caffeine contracture of intact muscle and the effect of caffeine on reticulum. J Gen Physiol 52: 750-759
37. Wendt IR, Stephenson DG (1983) Effects of caffeine on Ca-activated force production in skinned cardiac and skeletal muscle fibres of the rat. Pflugers Arch 398: 210-216

IMMUNOLOCALIZATION OF *CHARA* CALMODULIN AND THE REVERSIBILITY OF THE INHIBITION OF CYTOPLASMIC STREAMING BY Ca^{2+}

PETER P. JABLONSKY, RUTH P. HAGAN, FRANZ GROLIG, AND RICHARD E. WILLIAMSON

Plant Cell Biology Group, Research School of Biological Sciences, Australian National University, P.O. Box 471, Canberra, ACT 2601, Australia

Present address (F.G.): *Botanisches Institut I der Justus Liebig Universitat, Seckenbergstr. 17-21, D-6300 Giessen, Federal Republic of Germany*

Characean algal cells have been favored for studies of Ca^{2+}-regulated cytoplasmic streaming because of their giant size, simple structural organization, and possession of a readily elicited change in cytoplasmic-free Ca^{2+} concentration. The internodal cells are right cylinders with lengths of several centimeters and a diameter of approximately 0.5 to 1.0 mm. The stationary cortical cytoplasm underlying the plasma membrane contains most of the cell's microtubules and chloroplasts. The chloroplasts lie in files that trace a steep helix around the cell and each file of chloroplasts has about five bundles of F-actin filaments running parallel to it on the side of the chloroplasts facing the center of the cell. A layer of endoplasm, several micrometers thick, streams along these filament bundles at ≥ 50 $\mu m\ s^{-1}$ and is bounded by the tonoplast at its inner surface. Organelles and the actin-filament bundles react with an antibody to myosin (4). The membranous organelles of the streaming cytoplasm, but not those of the stationary cortex, therefore appear to be coated with myosin and, when ATP is depleted, endoplasmic organelles tightly bind the filament bundles (17). This is reminiscent of the rigor linkages formed between actin and myosin filaments after ATP depletion in muscle. It is therefore believed that organelle movements result from myosin binding to organelles and propelling them along stationary actin filaments.

Ca^{2+} concentrations >100 nM inhibit, to varying degrees, the ATP-dependent streaming in cells whose tonoplast (10, 12-14, 17) or plasma membrane (13) have been removed or permeabilized. Inhibition is complete at about 10^{-6} M Ca^{2+} in the most sensitive preparations, a concentration exceeded, *in vivo* (19), when the cell is stimulated to conduct an action potential. During such action potentials, streaming stops within a few hundred milliseconds of the time of maximum membrane depolarization (11) which, itself, is approximately coincident with the peak-free Ca^{2+} concentration (19).

Actin and myosin are both known to be regulated by Ca^{2+} through several different mechanisms. For example, myosin regulation occurs by Ca^{2+}-dependent phosphorylation of either its heavy (9) or light chains (1), or by the direct binding of Ca^{2+} to the light chains themselves (8); actin can be regulated in its ability to react with myosin, or through filament capping, cutting, or bundling proteins (7). The velocity at which plastic beads, coated with Ca^{2+}-stimulated (15) and Ca^{2+}-inhibited (5) myosins, move along characean actin bundles shows only the Ca^{2+}-sensitivity of the myosin used. This suggests that characean actin, itself, is not regulated and so, by default, requires the regulation of characean myosin. The Ca^{2+}-induced changes in myosin's distribution visualized by immunofluorescence (4) support this idea.

The ubiquitous Ca^{2+}-binding protein, calmodulin, regulates animal myosins by activating kinases phosphorylating the light (1) and heavy chains (9) of myosin II's, and by direct association with one member of the tailless myosin I category (6). As a result, calmodulin localizes in vertebrate cells to the stress fibers that contain actin and myosin II (3), and is an integral part of the actin-myosin I complex of microvilli. A rather different role for calmodulin in the regulation of characean myosin was proposed by Tominaga *et al.* (12, 14). Myosin inactivation, they proposed, results from phosphorylation catalyzed by a Ca^{2+}-dependent, calmodulin-independent protein kinase, while calmodulin activates a protein phosphatase that restores activity to myosin. The involvement of calmodulin was deduced only from the effects of calmodulin inhibitors, but a number of the familiar objections to their specificity are reduced when they are perfused inside a cell whose supply of Ca^{2+} and ATP is regulated artificially.

To further study calmodulin's role in the regulation of cytoplasmic streaming, we therefore examined whether it co-localizes with either actin or myosin. In the light of Tominaga's hypothesis (12, 14) that calmodulin is required to reverse the inhibition due to elevated Ca^{2+}, we also investigated whether it is retained in vacuole-perfused cells where the inhibitory effects of high Ca^{2+} are irreversible (18).

MATERIALS AND METHODS

Antibody BF8 is a monoclonal antibody raised by standard methods from mice immunized with calmodulin purified from pea leaves (Jablonsky *et al.*, submitted; 2). It is monospecific for calmodulin in a range of flowering plants except in the minority of species where it recognizes a protein with an apparent mol wt of 22.5 kD as well as calmodulin. Proteins precipitated from *Chara* extracts by TCA (4) were resolved by SDS PAGE and transferred to nitrocellulose (21). After fixation with 2% glutaraldehyde (16), immunoblotting with antibody BF8 was by the methods described in Williamson et al. (21).

Immunofluorescence of whole and perfused *Chara* cells was carried out with BF8 essentially as described by Grolig et al. (4), except that BSA was omitted from all solutions; glutaraldehyde-induced fluorescence was diminished by sodium borohydride reduction (1 mg ml^{-1} for 10 min following fixation and washing), and micrographs were taken on a Nikon Optiphot microscope using the standard filters for fluorescein and an additional red-excluding barrier filter. Perfusion solutions lacking ATP and containing 10^{-4} and 10^{-7} M free Ca^{2+} were used (AFPS and HC-AFPS of ref. 4).

RESULTS

A single band of 17.5 kD was specifically immunostained by antibody BF8 among the proteins precipitated by TCA (Fig. 1). When proteins were not rapidly precipitated in this way, one or more bands of lower M_r were immunostained, suggesting that TCA precipitation was inhibiting proteolysis.

The antibody was used for immunofluorescence of whole cells and of cells perfused to remove their tonoplast and much of the endoplasm. This leaves, *in situ*, the proteins required for the reactivation of Ca^{2+}-sensitive, ATP-dependent cytoplasmic streaming (17), but the inhibition is not reversed by lowering the Ca^{2+} concentration (18). In whole cells, there were substantial amounts of immunodetectable calmodulin in the endoplasm (Fig. 2), but very little in the cortex (not shown). Endoplasmic calmodulin was associated with a variety of organelles that were often aligned along what were presumed from their number, position, and orientation to be the actin bundles. Because antibodies are excluded from chloroplasts (20), and probably other membrane-bound organelles in cells prepared in perfusion solutions, the calmodulin is unlikely to be inside the organelles. The actin bundles, themselves, were not continuously stained so that the staining pattern resembled a string of beads (organelles) in which the string (actin) was largely unstained (Fig. 3). There was no calmodulin immunostaining at all in cells perfused with solutions containing either 10^{-7} or 10^{-4} M Ca^{2+} (Fig. 4).

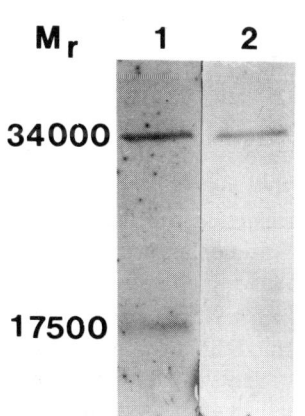

Figure 1. TCA-precipitated proteins from *Chara* immunoblotted with monoclonal anti-calmodulin. A single band of 17,500 M_r is specifically immunostained by BF8 (lane 1), and a band of 34,000 M_r is nonspecifically stained since it is still detected when BF8 is omitted (lane 2).

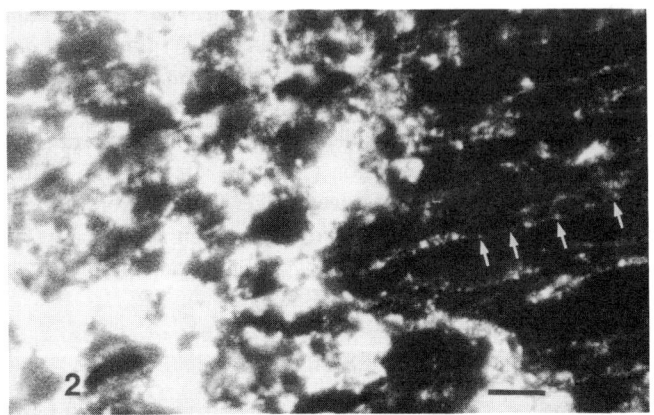

Figure 2. Calmodulin immunofluorescence of whole cells showing the strong labeling associated with the endoplasm. Where much of the endoplasm has been lost, fluorescence is associated with small spherical or ovoid organelles that are arranged in linear arrays (arrows), probably by binding to the actin bundles. The latter are not labeled. Bar = 5 μm.

DISCUSSION

Calmodulin is readily detected in whole cells by immunofluorescence and immunoblotting. It is widely distributed in the endoplasm, where it seems to associate with organelles that can bind to the actin filament bundles, but not with the actin bundles themselves. The cortex shows little labeling of organelles (including the chloroplasts which are adjacent to the calmodulin-rich endoplasm) and no labeling of microtubules. In the endoplasm, calmodulin resembles myosin in being associated with many organelles but, unlike myosin, is associated neither with endoplasmic strands (4) nor with the actin cables themselves. The distribution of calmodulin in whole cells will not reflect the protein's full distribution, *in vivo*, if soluble calmodulin extracts to some extent during preparation for microscopy. Extraction is suggested by the variation in the amount of immunostained endoplasm in different regions of a cell so that, in some areas, the actin bundles are overlain with abundant endoplasmic calmodulin, whereas in nearby areas, they are exposed.

The calmodulin detected in whole cells is, however, completely extracted by rapid vacuolar perfusion. Such perfusion is the first stage of the procedure to prepare a tonoplast-free cell in which Ca^{2+}-sensitive streaming can be reactivated by exogenous ATP (17). Perfused cells still contain actin (20) and myosin (4), showing that calmodulin is not tightly bound to either of the major force-generating proteins. This contrasts with the association of calmodulin with the stress fibers of vertebrate cells (3) and calmodulin's occurrence as a subunit of a member of the myosin I family (6). The velocity with which organelles move along the actin bundles when reactivated by ATP in *Chara*

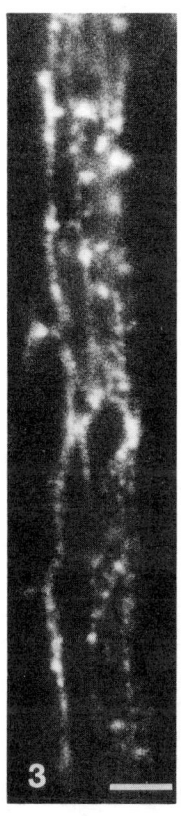

cells shows substantial Ca^{2+}-sensitivity (17), although probably somewhat less than exists *in vivo* (19). That this Ca^{2+}-sensitivity persists in the absence of immunodetectable calmodulin therefore strengthens the hypothesis (12, 14) that calmodulin does not inhibit the streaming at elevated Ca^{2+} concentrations. Similarly, the finding that the inhibition by high Ca^{2+} is not reversed when the Ca^{2+} concentration is lowered in such perfused cells (18) is consistent with the view of Tominaga *et al.* (12, 14) that calmodulin is instead required to reverse the inhibition of streaming.

In conclusion, the immunolocalization of calmodulin in whole and perfused internodal cells of *Chara* indicates that calmodulin is not firmly bound to either actin or myosin. Its absence from perfused cells in which organelle movements are irreversibly inhibited by elevated Ca^{2+} concentrations is consistent with the proposal of Tominaga *et al.* (12, 14) that calmodulin is required to restore myosin's activity, but not to inhibit it.

Figure 3. Calmodulin immunofluorescence showing the arrangement of the immunoreactive organelles along a group of 4 or 5 actin bundles that is associated with one file of chloroplasts. Bar = 5 μm.

Figure 4. Calmodulin immunofluorescence of a cell perfused with a solution containing 10^{-7} M Ca^{2+}. The photograph, taken by automatic exposure, required 3 to 4 times the exposure required for whole cells and shows only the weak chloroplast autofluorescence that passes the red-excluding filter. Identical results were obtained using 10^{-4} M Ca^{2+} in the solutions. Chloroplast autofluorescence is so weak as to be detectable only in the absence of immunostaining. Bar = 5 μm.

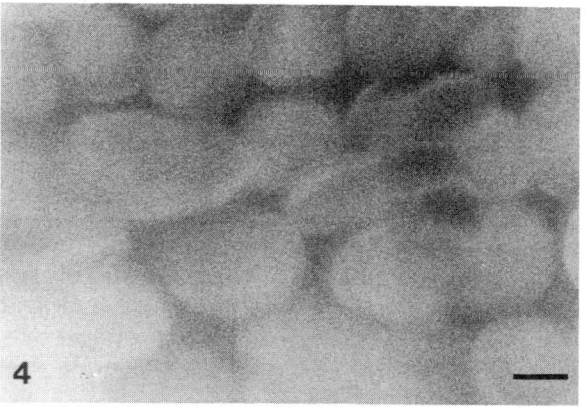

LITERATURE CITED

1. **Adelstein RS, Klee CB** (1981) Purification and characterization of smooth muscle myosin light chain kinase. J Biol Chem **256**: 7501-7509
2. **de Couet HG, Jablonsky PP, Perkin JL** (1986) Calmodulin associated with rhabdomeral photoreceptor microvilli of arthropods and squid. Cell Tissue Res **244**: 315-319
3. **Dedman JR, Welsh MJ, Means AR** (1978) Ca^{2+}-dependent regulator. Production and characterization of a monospecific antibody. J Biol Chem. **253**: 7515-7521
4. **Grolig F, Williamson RE, Parke J, Miller C, Anderton BH** (1988) Myosin and Ca^{2+}-sensitive streaming in the alga *Chara*: detection of two polypeptides reacting with a monoclonal anti-myosin and their localization in the streaming endoplasm. Eur J Cell Biol **47**: 22-31
5. **Kohama K, Shimmen T** (1985) Inhibitory Ca^{2+}-control of movement of beads coated with *Physarum* myosin along actin-cables in *Chara* internodal cells. Protoplasma **129**: 88-91
6. **Mooseker MS, Coleman TR** (1989) The 110-kD protein-calmodulin complex of the intestinal microvillus (Brush border myosin I) is a mechanoenzyme. J Cell Biol **108**: 2395-2400
7. **Pollard TD, Cooper JA** (1986) Actin and actin-binding proteins. A critical evaluation of mechanisms and functions. Annu Rev Biochem **55**: 987-1035
8. **Szent-Gyorgyi AG, Szentkiralyi EM, Kendrick-Jones J** (1973) The light chains of scallop myosin as regulatory subunits. J Mol Biol **74**: 179-203
9. **Tanaka E, Fukunaga K, Yamamoto H, Iwasa T, Miyamato E** (1986) Regulation of the actin-activated Mg-ATPase of brain myosin via phosphorylation by the brain Ca^{2+}, calmodulin-dependent protein kinases. J Neurochem **47**: 254-262
10. **Tazawa M, Kikuyama M, Shimmen T** (1976) Electric characteristics and cytoplasmic streaming of Characeae cells lacking tonoplast. Cell Struct Funct **1**: 165-176
11. **Tazawa M, Kishimoto U** (1968) Cessation of cytoplasmic streaming of *Chara* internodes during action potential. Plant Cell Physiol **9**: 361-368
12. **Tominaga Y, Muto S, Shimmen T, Tazawa M** (1985) Calmodulin and Ca^{2+}-controlled cytoplasmic streaming in characean cells. Cell Struct Funct **10**: 315-325
13. **Tominaga Y, Shimmen T, Tazawa M** (1983) Control of cytoplasmic streaming by extracellular Ca^{2+} in permeabilized *Nitella* cells. Protoplasma **116**: 75-77
14. **Tominaga Y, Wayne R, Tung HYL, Tazawa M** (1987) Phosphorylation-dephosphorylation is involved in Ca^{2+}-controlled cytoplasmic streaming of characean cells. Protoplasma **136**: 161-169
15. **Vale RD, Szent-Gyorgyi AG, Sheetz MP** (1984) Movement of scallop myosin on *Nitella* actin filaments: regulation by calcium. Proc Natl Acad Sci USA **81**: 6775-6778
16. **Van Eldik LJ, Wolchok SR** (1984) Conditions for reproducible detection of calmodulin and S100 in immunoblots. Biochem Biophys Res Commun **124**: 752-759
17. **Williamson RE** (1975) Cytoplasmic streaming in *Chara*: a cell model activated by ATP and inhibited by cytochalasin B. J Cell Sci **17**: 655-668
18. **Williamson RE** (1979) Filaments associated with the endoplasmic reticulum in the streaming cytoplasm of *Chara corallina*. Eur J Cell Biol **20**: 177-183

19. **Williamson RE, Ashley CC** (1982) Free Ca^{2+} and cytoplasmic streaming in the alga *Chara*. Nature **296**: 647-651
20. **Williamson RE, McCurdy DW, Hurley UA, Perkin JL** (1987) Actin of *Chara* giant internodal cells. Plant Physiol **85**: 268-272
21. **Williamson RE, Perkin JL, McCurdy DW, Craig S, Hurley UA** (1986) Production and use of monoclonal antibodies to study the cytoskeleton and other components of the cortical cytoplasm of *Chara*. Eur J Cell Biol **41**: 1-8

CALCIUM, CYTOPLASMIC STREAMING, AND GRAVITY

RANDY WAYNE, MARK STAVES, AND YUJI MORIYASU

Section of Plant Biology, Cornell University, Ithaca, NY 14853 USA

It is fascinating to consider the mechanisms involved in the perception of gravity in plants. While sedimenting statoliths, including amyloplasts and $BaSO_4$ crystals, are considered the most likely candidates for the gravisensor, there are a number of gravity-sensing systems that do not have statoliths. In these cases, the whole cell may act as the gravisensor and the plasma membrane of the walled cells may act as the gravireceptor, through the compression and/or the tension it experiences in a gravitational field. While it is likely that many different gravireceptors have evolved for the sensory detection of gravity in varied ecological environments, and in each evolutionarily distinct organism, it is also possible that the plasma membrane is a common gravireceptor in all organisms, with statoliths acting as antennae that increase the sensitivity of the responding cells to gravity. We have chosen the internodal cells of *Nitellopsis* as a model system to study graviperception because it is devoid of any visible statoliths.

The force of gravity is extremely weak and it is likely that the gravitational energy is transduced into a biophysical and/or biochemical form of energy that can be amplified. The calcium ion is a universal component of the amplification mechanisms involved in coupling stimuli to responses. Ca^{2+} is required for the gravitropic reactions of the roots of higher plants. We are using the gravity-regulated polarity of cytoplasmic streaming as a tool to understand graviperception and signal transduction in plants. Here, we present our results that support the hypothesis that the plasma membrane is the gravireceptor and that Ca^{2+} ions participate in the signal transduction chain that leads to a gravity-induced polarity of cytoplasmic streaming.

MATERIALS AND METHODS

Nitellopsis obtusa (Desv. in Lois.) J. Gr. and *Chara corallina* Klein ex Willd., em R.D.W. (= *Chara australis* R. Brown) were grown in a soil-water mixture in large glass tanks at 28°C with continuous light. Typically, young internodal cells of *Nitellopsis* or *Chara* (2-4 cm) were isolated and placed in artificial pond water (APW) (0.1 mM NaCl, 0.1 mM KCl, and 0.1 mM $CaCl_2$)

buffered by 2 mM Hepes-NaOH, pH 7.2, at least overnight before use. Sometimes cells that had a buildup of $CaCO_3$ crystals on the walls were used. In this case, the cells were first placed in APW buffered by 10 mM Mes titrated with Tris to pH 5.5. The cells remained in this medium overnight and were then transferred to APW buffered with 2 mM Hepes-NaOH, pH 7.2. The cells remained in this medium at least overnight before use. All experiments were carried out at $25 \pm 2°C$.

Gravity Experiments on *Nitellopsis*

For the gravity experiments, cells were placed in a Plexiglas chamber in which the cell ends were embedded in Dow Corning silicone grease. The rest of the cell was bathed in APW buffered with 2 mM Hepes-NaOH with, or without, inhibitors. The chamber was covered with a cover glass and then placed on a Zeiss Standard horizontal microscope equipped with a rotating stage. The optics of the Zeiss horizontal microscope included a Plan 16 X objective (NA = 0.35), a 2x optivar, and Kpl-W 10x/18 oculars.

The cells were oriented so that the two opposing streams adjacent to the indifferent zone could be observed simultaneously. Streaming in an area midway along the length of the internode was observed. The velocity of the streaming endoplasm 6 to 12 (depending on cell geometry) chloroplast rows away from the indifferent zone was measured. Unless stated otherwise, the velocities of the particles in the bulk endoplasm adjacent to the actin bundles were measured. This was accomplished, under our optical conditions, when the chloroplasts were in sharp focus. The cells typically were oriented so that the actin bundles were oriented parallel to the force of gravity.

The time it took for small endoplasmic particles to flow across a 137 μm grid was measured with a stopwatch. Twenty-five measurements were made on each side. The polar ratio is defined as the ratio of the velocity of the downwardly-directed stream to the velocity of the upwardly-directed stream. Details of the methods can be found in Wayne *et al.* (4).

Ca^{2+}-EGTA buffers were made with the aid of a computer program. All the buffers included 1 mM EGTA (titrated with NaOH to pH 7.2), 2 mM Hepes (titrated with NaOH to pH 7.2), 0.1 mM NaCl, 0.1 mM KCl, 1 mM free Mg^{2+} (added as $MgCl_2$), and various concentrations of $CaCl_2$.

Excitation-Contraction Coupling Experiments on *Chara*

Cytoplasmic streaming was observed with bright-field optics on either an Olympus BH-2 microscope equipped with a 10 X SPLAN objective (NA = 0.3) connected to a NC-8 CCD Video camera and a PM1971A autocolor monitor (NEC, Tokyo, Japan) or a Zeiss IM35 inverted microscope equipped with a 10 X objective (NA = 0.25) and Kpl- W10X/18 oculars. The cells were stimulated with a S-95 Tri-Level Stimulator (Medical Systems Corp., Greenvale, NY) by applying 1.5 V for a duration of 0.1 s through Ag/AgCl wires placed parallel to the cell in two electrically isolated chambers. The cells received approximately 1.1 μA of current. The cells were incubated in similar Ca^{2+}-EGTA buffers as described above, except the $MgCl_2$ and NaCl were not included in the experiments with *Chara*.

Membrane potentials (E_m) were measured with microcapillary electrodes that were placed in the cytoplasm. The electrodes were positioned with a Narashige MO-103 Hydraulic micromanipulator. The microcapillary electrodes were pulled from Narashige GD-1 capillaries on a Narashige PP-83 Pipette Puller (Narashige Sci. Instr., Tokyo, Japan) and filled with 3 M KCl. The reference electrode was made by filling a plastic pipette with 2% agar dissolved in 100 mM KCl. The potentials were amplified with a S-7071A electrometer equipped with a tangerine probe (WPI, Quinnipiac, CT) and recorded on a Model 2125M strip-chart recorder (Allen Datagraph, Inc., Salem, NH).

RESULTS AND DISCUSSION

In a horizontally placed internodal cell of *Nitellopsis obtusa*, the velocity of cytoplasmic streaming adjacent to the indifferent zone in one direction is equal to the velocity of cytoplasmic streaming in the opposite direction. However, if the cell is rotated 90 degrees, the endoplasm in the downwardly-directed stream flows approximately 10% faster than the endoplasm in the upwardly-directed stream. This leads to a polarity in streaming, which we define as the polar ratio. The average polar ratio of vertically-positioned cells is 1.110 ± 0.006 (n=51) compared to 1.009 ± 0.003 (n=28) for horizontally-positioned cells.

We have begun efforts to localize the region of the cell involved in graviperception and/or the early events of signal transduction. Using cellular ligation, we have been able to remove both of the cell ends to test whether the two cell ends, which include nodal cells and plasmodesmata, are necessary for the response. We find that cell fragments from which both cell ends have been removed, although streaming normally, cannot respond to gravity. We also find that internodal cells that possess only one node still are not able to respond to gravity. Cells that possess only their distal or proximal node do not respond to gravity irrespective of their orientation. Furthermore, treatment of a cell end with 31.2 kJm^{-2} of UV light completely eliminates the gravity-induced polarity of cytoplasmic streaming. By contrast, treatment of the middle of the cell with UV light has no effect on the gravity-induced polarity. These data indicate that the cell ends are specialized regions that participate in the response to gravity. We postulate that the plasma membrane or the plasmodesmata at the cell ends are involved in graviperception and/or the early events of signal transduction.

Since the cell ends appear to be required for the graviresponse, we tested the possibility that the falling of the protoplasm *in toto* is required for the realization of the graviresponse. The static bouyancy of the protoplasm of a given cell can be varied by changing the external medium. When a cell is placed in APW, its bouyancy is -1.18 μN. By contrast, the same cell's bouyancy is

2.88 μN when it is placed in APW containing 10% BSA. The whole protoplasm within the wall thus sinks in APW and floats in APW containing 10% BSA. When the cells are placed in APW without BSA, the polar ratio is greater than 1. The polar ratio declines to 1 as the BSA concentration increases and eventually becomes less than 1 at high BSA concentrations. These data indicate that the cells of *Nitellopsis* are capable of sensing the weight of the protoplast. The falling of the protoplast *in toto* within the cell wall may cause a deformation of the top and bottom plasma membrane which may lead to the graviresponse.

Further evidence that the membrane may be involved in graviperception, or the early stages of signal transduction, comes from experiments where we found that staining the cell with neutral red results in a reversal of the gravity response. Cells stained with neutral red stream up faster than they stream down. The effect of neutral red on the polar ratio is concentration-dependent and reversible. Since neutral red is known to activate a K^+ channel and cause a membrane hyperpolarization upon illumination (1), we tested the effect of the voltage-dependent K^+ channel blocker, tetraethylammonium Cl^- (TEA) on cells treated with neutral red and light. TEA prevents neutral red from reversing the gravity effect and, consequently, vertically positioned cells show a polar ratio greater than 1 whether the cells are given neutral red or not. The effect of TEA is concentration-dependent. These data indicate that a hyperpolarization may be correlated with a reversed response to gravity. Perhaps this means that a depolarization may be correlated with a normal response to gravity. Indeed, treating the cells with 1 mM KCl, which depolarizes the membrane, enhances the responsiveness of *Nitellopsis* to gravity. Treatment of the cells with concentrations of KCl that will completely depolarize the membrane results in a reversed response to gravity.

Since a depolarization is often correlated with the opening of voltage-dependent Ca^{2+}-channels, we treated the cells with the voltage-dependent Ca^{2+}-channel blocker nifedipine to determine whether or not Ca^{2+} influx through voltage-dependent channels is required for the graviresponse. Nifedipine inhibits the response in a concentration-dependent and reversible manner. Interestingly, nifedipine inhibition saturates at a polar ratio of 1. $LaCl_3$, a very potent, but irreversible, inhibitor of the Ca^{2+} current in Characean cells, causes a reversed response to gravity.

Cells treated with Ca^{2+}-EGTA buffers reveal a Ca^{2+}-dependence in their response to gravity. Cells treated with Ca^{2+} concentrations lower than 1 μM show a reversal in the normal response to gravity and thus have a polar ratio less than 1. Cells treated with Ca^{2+} concentrations higher than 1 μM exhibit a normal response to gravity and a polar ratio greater than 1. The response to gravity is supported half-maximally by 0.91 μM Ca^{2+}. The polarity of horizontally placed cells is independent of external Ca^{2+}. The response to gravity requires Ca^{2+} specifically as a bivalent cation. The order of effectiveness of other bivalent cations follows the sequence: $Ca^{2+} >> Sr^{2+} > Mg^{2+} > Ba^{2+}$.

We find that the streaming velocity of intact horizontal cells is Ca^{2+} dependent. The optimum concentration of Ca^{2+} is 1 μM; the threshold is 0.1 μM, and 0.4 μM supports a half-maximal response on the ascending portion of the curve. These data indicate that either lowering or raising the $[Ca^{2+}]$ around 0.4 μM will slow down or speed up cytoplasmic streaming. Gravity may induce a very slight increase in the intracellular $[Ca^{2+}]$ in the downwardly-directed stream, and a very slight decrease in the $[Ca^{2+}]$ in the upwardly-directed stream.

The Ca^{2+} concentrations required to activate the gravity response are in the range of physiological concentrations that activate calmodulin. In order to test the possibility that the Ca^{2+}-calmodulin complex participates in the gravity-induced signal-transduction chain, we treated the cells with the Ca^{2+}-calmodulin complex inhibitors, W-7 (50 μM) and R 24571 (2 μM). Both drugs show complex inhibition kinetics. Both drugs cause a reversal of the polarity within 5 min after treatment. This is followed by a return to the normal polarity within 15 min, and then an enhanced polarity. The degree of enhancement depends upon the drug concentration. The effect of these inhibitors may be nonspecific since 50 μM W-5 yields similar results. The reversal of polarity cannot be obtained again by re-treating the cells with fresh inhibitor.

We propose that gravity induces a falling of the protoplast within the cell wall. The plasma membrane may then be compressed against the wall on the bottom and come under tension on the top of the cell. These deformations then modulate various channels, and/or transport proteins that may create a transcellular electric field. The two cell ends may then communicate with each other by means of this electric field. This field may either be created through the action of Ca^{2+} or, in turn, it may help to redistribute Ca^{2+} in the upwardly- and downwardly-directed stream and thus regulate the polarity of cytoplasmic streaming.

In order to investigate the mechanism of Ca^{2+} action on the gravity-induced polarity of cytoplasmic streaming, we tested the effect of gravity on another genus of the Characeae, *Chara*. Like *Nitellopsis*, *Chara corallina* is also responsive to gravity. The polar ratio is 1.000 in horizontal cells and 1.115 ± 0.021 (n=4) in vertically-oriented cells.

The effect of an action potential on cytoplasmic streaming has become a classic example in studying the effect of Ca^{2+} on stimulus-response coupling. Through the work of Williamson's and Tazawa's groups, we know that an action potential induces an increase in the cytoplasmic-free Ca^{2+} concentration and that concentrations of Ca^{2+} greater than 1 μM will stop streaming when applied to tonoplast-free or permeabilized cells (2, 3, 5). While these studies focused on the effect of intracellular Ca^{2+} on the cessation of cytoplasmic streaming in cell models, we tested the effect of extracellular Ca^{2+} on the cessation of streaming in response to an action potential (E-C coupling). Intact cells subjected to various concentrations of extracellular Ca^{2+} show the identical sensitivity to Ca^{2+} in their cessation of streaming in response to an action potential as do cell models in response to changes in the intracellular Ca^{2+} concentration. The equivalent sensitivities to Ca^{2+} in the two systems is curious since the electrochemical potential for the influx of Ca^{2+} at pCa 6 is 88% of the electrochemical potential for the influx of Ca^{2+} at pCa 5 if the membrane potential is assumed to remain constant at -0.186 V.

Why then, do the intact cells show the same sensitivity to Ca^{2+} as the cell models? Using Ca^{2+}-EGTA buffers, we find that the membrane potential is sensitive to the concentration of external Ca^{2+}. The membrane is hyperpolarized at pCa 4 and pCa 5, but becomes depolarized in approximately 5 min upon being transferred to a medium buffered at pCa 6, pCa 7 or pCa 8. Therefore, the driving force for Ca^{2+} influx, which is the difference between the membrane potential and the Nernst potential for Ca^{2+}, is -0.267 V and -0.107 V, for pCa 5 and pCa 6, respectively. That is, the driving force for Ca^{2+} influx when the extracellular Ca^{2+} concentration is buffered at pCa 6 is less than half of what it is when the extracellular medium is buffered at pCa 5. Therefore, the rate of Ca^{2+} influx may determine the ability of the cell to respond to an action potential.

The rate of Ca^{2+} influx depends both upon the electrochemical potential and the Ca^{2+} conductance of the plasma membrane. It is becoming established through pharmacological experiments that both voltage-dependent Ca^{2+} channels, whose conductance increases upon membrane depolarization, and voltage-independent channels exist in plant cells. Although we originally believed that an action potential opens voltage-dependent Ca^{2+} channels, our new data, which show that the ability of a cell to undergo E-C coupling is correlated with a hyperpolarized membrane, have caused us to rethink this question. Indeed, we now think that voltage-independent Ca^{2+} channels may participate in E-C coupling in Characean cells. This suggestion is supported by the observation that nifedipine (250 μM, 5-90 min), does not inhibit E-C coupling in *Chara*, even though it is an effective blocker of a number of physiological responses in characean cells.

In order to further characterize the effect of Ca^{2+} on the membrane potential, we treated the cells with the chloride channel blocker, 9-Anthracenecarboxylic acid (A-9-C). While the cells depolarize upon low Ca^{2+} treatments in the presence of 1 mM A-9-C, the rate of membrane depolarization in response to low Ca^{2+} concentrations is reduced by approximately 3 to 4 times. Therefore, it appears that the permeability to Cl⁻ increases in the presence 1 μM or less external Ca^{2+} and this leads to a depolarized membrane. By contrast, the permeability of the membrane to Cl⁻ may decrease at high external Ca^{2+} concentrations and this results in a hyperpolarized membrane.

When we treat cells buffered at pCa 6 or pCa 7 with A-9-C, we have a window of approximately 20 min before the membrane becomes depolarized to the control (pCa 6 or pCa 7) values. Therefore, we can test the ability of the cells to respond to pCa 6 or pCa 7 when the membrane is in the hyperpolarized state. Whereas none of the cells treated with pCa 6 or pCa 7 show E-C coupling after 7 min in the buffer, all the cells treated with A-9-C show E-C coupling at 7 min. These data show that the concentration of external Ca^{2+} that will support a half-maximal response depends on the membrane potential. In this case, the concentration of external Ca^{2+} that supports a half-maximal response can be shifted at least two orders of magnitude by manipulating the membrane potential.

Our observations show that the ability of Ca^{2+} to couple the stimulus to the response during E-C coupling depends on the electrochemical potential for Ca^{2+} and that the effect of external Ca^{2+} on the membrane potential plays a more prominent role than the effect of external Ca^{2+} on the chemical potential in establishing the electrochemical potential for Ca^{2+}. That is, external Ca^{2+} itself can effectively control its own electrochemical potential by influencing the membrane potential.

Now, in terms of the role of Ca^{2+} in influencing the gravity-dependent polarity of cytoplasmic streaming, we must not only consider the role of membrane potential in regulating Ca^{2+} fluxes, but also the role of Ca^{2+} in regulating the membrane potential.

ACKNOWLEDGMENTS

We would like to thank A. Carl Leopold and Eiji Samitsubo for their interest. This work was supported, in part, from grants from NASA (NA 610-0071) and USDA (185-6402) to R.W.

LITERATURE CITED

1. **Tazawa M, Shimmen T** (1980) Demonstration of the K^+ channel in the plasmalemma of tonoplast-free cells of *Chara australis*. Plant Cell Physiol **21**: 1535-1540
2. **Tominaga Y, Shimmen T, Tazawa M** (1983) Control of cytoplasmic streaming by extracellular Ca^{2+} in permeabilized *Nitella* cells. Protoplasma **116**: 75-77
3. **Tominaga Y, Wayne R, Tung HYL, Tazawa M** (1987) Phosphorylation-dephosphorylation is involved in Ca^{2+}-controlled cytoplasmic streaming of Characean cells. Protoplasma **136**: 161-169
4. **Wayne R, Staves M, Leopold AC** (1990) Gravity-dependent polarity of cytoplasmic streaming in *Nitellopsis*. Protoplasma (in press)
5. **Williamson RE, Ashley CC** (1982) Free Ca^{2+} and cytoplasmic streaming in alga *Chara*. Nature **296**: 647-651

CALCIUM AND THE REGULATION OF MITOSIS

Peter K. Hepler, Dahong Zhang, and Dale A. Callaham

*Department of Botany, University of Massachusetts,
Amherst, MA 01002, USA*

Calcium ions (Ca) are widely recognized as possible regulators of the events of mitosis (9, 13, 17, 18, 36). Ever since the discovery by Weisenberg (33) that elevated levels of Ca promoted the depolymerization of microtubules (MTs), and conversely, low levels of the ion permitted their polymerization, it has been attractive to imagine that fluctuations in the [Ca] would control the structure and, possibly, the function of the mitotic apparatus. But, beyond the obvious association of Ca with MT formation, the ion could affect a variety of other processes, such as the activation of a mechanochemical motor (*e.g.* dynein), the sol-gel transitions of spindle-associated microfilament networks, the transformations of the spindle-associated membrane system, and the condensation and maturation of the chromosomes, and thereby contribute to the control of mitosis (13, 18).

It is the purpose of this chapter to review our recent studies on the Ca-mitosis problem. In the sections that follow, we first briefly discuss aspects of the structure of the mitotic apparatus that relate to the Ca question. We then discuss recent evidence concerning the occurrence and timing of endogenous Ca transients in the mitotic apparatus. We follow this with a summary of our physiological studies in which attempts have been made to modulate both internal and external Ca and thereby modulate the events of mitosis. Finally, we attempt to bring these results together to emphasize what is known, but also to direct attention to unknown and fruitful areas for future research.

STRUCTURAL FEATURES OF THE MITOTIC APPARATUS

MTs are common elements of all mitotic apparatuses; by some mechanism, they contribute to the motion of chromosomes to opposite poles during mitosis (23). Although not so widely recognized, membrane elements are also a common structural feature of the mitotic apparatus (11, 14, 17, 18). Even though the nuclear envelope breaks down and disperses in higher eukaryotic cells, the emerging mitotic apparatus may still be surrounded by an extensive endomembrane system composed largely of ER. Of particular interest are the observations from different species showing that elements, composed largely of smooth, tubular ER, extend from the spindle pole into the mitotic apparatus interior, specifically along kinetochore MTs (11, 14, 18). Thus, spindle membranes display a close structural association with the MTs and, in addition, undergo transformations in concert with the mitotic cycle.

These spindle membranes seem important since they may participate in the regulation of Ca within the mitotic apparatus (9, 14, 18). It has been attractive to compare them to the sarcoplasmic reticulum (SR) of muscle, a membrane system well known for its role in the release and resequestration of Ca that is essential for muscle contraction. Like the SR, the spindle-associated ER is physically close to the motile macromolecules that it might control. Indeed, in dividing barley leaf cells, the tubular, reticulate ER intimately interdigitates with the kinetochore MTs and extends completely to the point where the MTs insert on the chromosome (11). Studies in different animal cell systems have shown that these membranes, like those of the SR, contain a Ca-ATPase (for review, 18) and also a calsequestrin-like protein (10). Our own efforts toward supporting the contention that these membranes control Ca have come from our studies both at the light and electron microscope level in which we have attempted to spatially localize the ion. Ultrastructural investigations using pyroantimonate to complex exchangeably bound Ca reveal the presence of electron-dense deposits in the spindle-associated membranes of dividing barley leaf cells and spermatocytes of *Marsilea* (34). Subsequent studies with the vital Ca chelate probe chlortetracycline (CTC), which localizes membrane-associated Ca, delineates "cones" of fluorescence within the mitotic apparatus of living endosperm cells of *Haemanthus* that exactly overlay the birefringent kinetochore fibers (39, 40).

Attempts to quantitate possible changes in the level of the membrane-associated Ca provided preliminary evidence for its reduction a few minutes prior to the onset of anaphase (41). Inspection of the morphology also indicated that the "cones" of fluorescence tended to disappear by late metaphase and were not observed at any stage of anaphase (41). Parallel studies using two permeant, voltage-sensitive dyes, dioxacarbocycanine and anilino naphthalene sulfonate, provided additional information that ion changes, possibly involving Ca, occurred in the mitotic apparatus at the time of the metaphase/anaphase transition, raising the tantalizing possibility that these ion changes might be regulatory (41) (Fig. 1).

PHYSIOLOGICAL STUDIES

Determination of Changes in Intracellular-Free Ca

Of all the questions surrounding the Ca-mitosis problem, one of the most important concerns the issue of whether changes in the intracellular or intraspindle [Ca] occur and, if so, when, where, and to what magnitude (13). We have spent considerable effort on this problem and have yet to generate results with which we are entirely satisfied. Without laboring over the details, suffice

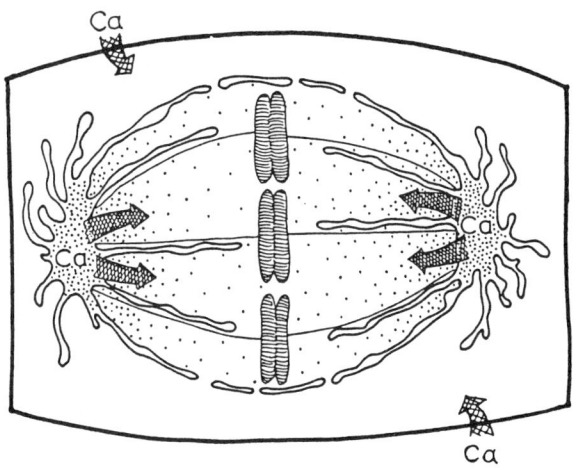

Figure 1. A diagrammatic representation of a cell in late metaphase showing the spindle apparatus and associated membranes. It is suggested that calcium influx is regulated both at the plasmalemma and from the endomembranes. Figure reproduced from Hepler (12) with permission from *The Journal of Cell Biology*.

to say that these measurements are extremely difficult to make because the resting [Ca] is very low (0.1-0.2 μM), and because a single cell, or more precisely its mitotic apparatus, presents a very small target from which to collect relevant optical signals reflecting ion transients (3). Also, some of the favored Ca probes, *e.g.* the fluorescent indicators (8), have a tendency to compartmentalize in the vacuole and become relatively useless for intracytoplasmic Ca measurements (3). Nevertheless, we have had some success using differential microspectroscopy to measure small changes in absorbance in mitotic cells loaded by microinjection with the metallochromic dye arsenazo-III (15). These results are summarized below.

Our results show that an increase in [Ca] occurs during mitosis, but that the rise begins after the chromosomes have undergone their initial splitting (15) (Fig. 2). The [Ca] increases gradually as the chromosomes move to the poles, and then after reaching a plateau, gradually declines, returning to basal levels at about the time the new cell plate first appears. Because the rise begins after the actual onset of anaphase, it would seem that it cannot be the trigger for the metaphase/anaphase transition. However, the rise in [Ca] during anaphase supports the view that the ion may participate in the regulation of chromosome motion (15). The data, for example, are entirely consistent with the idea that elevated levels of Ca facilitate the depolymerization of MTs, a process which must occur in order for the chromosomes to approach the poles. Also, the subsequent decline in the [Ca] during late anaphase, early telophase, fits well with the presumed need to have low [Ca] in order to support MT assembly in the phragmoplast (15) (Fig. 2).

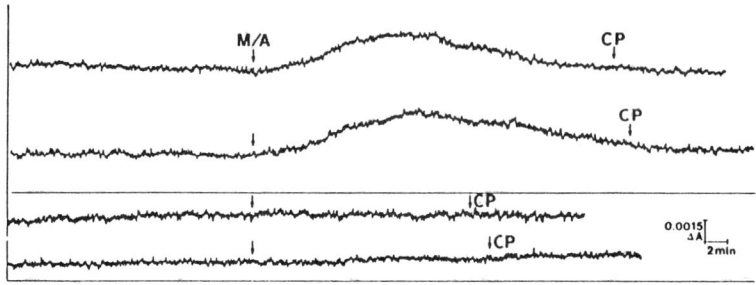

Figure 2. Metaphase through telophase in stamen hair cells of *Tradescantia* that have been loaded with the absorbance indicator dye, arsenazo-III. A gradual increase in the [Ca] is observed to begin immediately after the metaphase/anaphase transition (M/A), and continues while the chromosomes move to the poles. Thereafter, the Ca gradually declines, reaching the basal levels at the time of cell plate formation (CP). The top two traces show two separate dye-injected cells while the lower two traces are from dividing cells lacking dye. Figure reproduced from Hepler and Callaham (15) with permission from *The Journal of Cell Biology*.

At first, our results showing a gradual increase in [Ca] during anaphase appeared to be at odds with the published literature from studies on different animal cell systems. Poenie *et al.* (27) reported the occurrence of a brief (20 s) spike in [Ca] at the metaphase/anaphase transition in cultured PtK cells. However, others have failed to repeat these studies and find, instead, that the spikes can occur at any point during mitosis (28, 31) and, even more importantly, that they are due to the presence of fetal calf serum in the medium (1, 31). Cells grown in the absence of serum can be cultured through mitosis without displaying any spikes in the [Ca]. Two studies have, however, reported the occurrence of gradual increases in the [Ca] that begin during mid-late metaphase and extend through anaphase (28, 31). Although not identical to that which we have reported for *Tradescantia*, nevertheless, there is considerable similarity in showing an elevated level of Ca during anaphase. The most compelling example of a regulatory pulse of Ca has been found at the prophase/prometaphase transition in sea urchins, where it precedes and appears necessary for nuclear envelope breakdown (30).

Still, the whole issue of the presence, timing, magnitude, and location of Ca transients during mitosis cannot be considered as being solved. There are enormous technical problems in working with the various Ca indicators and some of the published reports, including our own, may well suffer from these measuring difficulties that have little or nothing to do with the what is going on inside the cell (3). In our studies using arsenazo-III, for example, there is insufficient sensitivity to detect Ca changes below 0.3 μM, and thus, we may have missed some small, but important, transients (15). An even more important consideration may be the fact that the crucial regulatory pulses might be confined to localized spatial domains (1), as defined by the highly structured intraspindle membrane system (18), described previously. It is possible to imagine that substantial changes in the [Ca] might have occurred, but because they were spatially confined, they would have had little impact on the global, pan-cellular [Ca]. In our studies on *Tradescantia*, we collected a signal from the entire mitotic apparatus area and could have missed local changes (15). Other studies that have used sensitive video equipment to image, and thus, spatially localize the changes in [Ca], have not yet provided compelling evidence for highly localized changes (27, 29). But again, one cannot consider the problem as solved since further technical difficulties surround the problem of spatial imaging of low-level light signals. Continued effort on the determination of the occurrence, timing, magnitude, and location of Ca transients is essential if we are to understand the role of this ion in the events of mitosis.

Modulation of Extracellular Ca

One approach to deciphering the role of Ca during mitosis has been to modulate the level of the ion in the extracellular medium or to apply agents such as lanthanum or Ca-channel blockers to prevent the ion from entering the cell and then ask what effect these treatments have on the timing and the events of mitosis (12, 36). Using dividing stamen hair cells of *Tradescantia*, we have examined the effect that culture in very low [Ca] has on the events and progress of mitosis (12). The appropriate culture conditions are established with EGTA buffers, in which the [Ca] is below 0.1 μM, while the [Mg], by contrast, is held at 1 mM or close to its normal level. The results clearly show that low extracellular Ca greatly prolongs the cells' progression through metaphase and retards the entry into anaphase (12). Interestingly though, when the cells enter anaphase, they complete the remaining stages of mitosis, anaphase motion and cell plate formation, in normal time. Culture of the cells in either lanthanum or the Ca-channel blocker, methoxy-verapamil, also provides evidence that the metaphase/anaphase transition appears to be particularly sensitive to Ca restriction (12).

Further studies on *Tradescantia* stamen hair cells by Wolniak and co-workers (4, 22, 35, 37, 38) provide additional support for the idea that the metaphase/anaphase transition requires Ca. Using quin2 as an extracellular Ca-chelator, they, too, have momentarily arrested cells in metaphase, but then, through the addition of Ca or Ca plus the ionophore A-23187, they have been able to overcome the block and induce cells to enter anaphase (37). They have also applied other Ca channel blockers, including nifedipine and diltiazem, and succeeded in arresting cells in metaphase (38), observations that agree well with our own using methoxy-verapamil. Taking the studies a step further, they have made use of the fact that nifedipine is photoinactivatable with brief pulses of UV light, and have shown that the metaphase block can be overcome by irradiation, especially in the presence of 100 μM external Ca (38). Finally, Chen and Wolniak (4) have used a Ca channel agonist, BAY K-8644, and reported that metaphase transit time can be substantially accelerated if applied shortly after nuclear envelope breakdown.

It seems reasonable to conclude from these studies that a pulse of Ca is required during late metaphase to induce cells to enter anaphase, but that once that threshold has been achieved, the cells are capable of completing mitosis without further input of extracellular Ca. A model, shown in Figure 1, depicts a late metaphase cell in which the chromosomes are poised to undergo their initial splitting. The suggestion is made here that an influx of Ca, possibly through voltage-gated channels, stimulates the further release of Ca that is stored in the intraspindle ER system and that, together, these ion changes promote the rapid, synchronous splitting of the sister chromatids and facilitate their motion to opposite poles. However, the results from the direct measurement of intracellular Ca fail to show a rise at the metaphase/anaphase transition and, thus, do not support the concept of "trigger pulses" as a global phenomenon (15). There may be a rise in Ca during anaphase that regulates aspects of anaphase chromosome motion, but the actual stimulus that initiates anaphase would seem to be in question. In light of this conflict, it is pertinent to mention recent observations by Larsen *et al.* (22) showing that the quin2 arrest in metaphase can be overcome by the permeant diacylglycerol analog, 1,2 dioctanoylglycerol, but not by the inactive analog 1,3 dioctanoylglycerol. These results suggest that protein kinase-C activation may be important for completing the metaphase/anaphase transition (22). Since C-kinase requires Ca, but not at an elevated level (2, 26), it might help explain the current paradox in which the observations show that Ca can contribute to the events that cause anaphase onset, but that a so-called "trigger pulse" on a global scale has not been established.

Modulation of Intracellular Ca

While our understanding of mitotic regulation has benefited from experiments that modulate the extracellular Ca milieu, more specific information can be gained through the modulation of the ion within the cytoplasm. On several occasions, we have attempted to use the ionophore, A-23187, together with buffered solutions of differing [Ca], but without much success. Mitotic perturbations were always variable, and we also have been unable to cause the well-known Ca-inhibition of cytoplasmic streaming (16) when we applied A-23187 and high Ca to actively streaming, but nondividing, stamen hair cells. We have, instead, taken the direct approach of injecting either Ca or Ca buffers into mitotic cells at different phases and monitored their effect. Coupled with these studies have been parallel studies on the effect that these injections have on the intracellular level of Ca, as measured using the fluorescent indicator, indo-1, rather than arsenazo-III. Despite the vacuolar compartmentation exhibited by indo-1, the process generally takes over 30 min, and thus, problems can be largely eliminated by conducting experiments within a short time period ($<$ 20 min). A major advantage for indo-1 is its markedly greater sensitivity to changes in [Ca] close to the resting level when compared to that for arsenazo-III (3, 8). The work is summarized below.

We have performed a series of studies in which a transient pulse of Ca has been directly injected into the mitotic apparatus at different times after nuclear envelope breakdown and before anaphase onset in an effort to see whether metaphase transit time could be experimentally accelerated or retarded (Hepler, unpublished observations). The Ca was injected ionophoretically to approximately 1 μM, a level that will just inhibit cytoplasmic streaming if applied to a nondividing cell. Preliminary observations indicate that these 1-μM pulses of Ca neither retard nor accelerate metaphase transit time when applied at 10- to 20-min post-nuclear-envelope breakdown. Similar studies on cultured animal (PtK) cells have reported that Ca injection causes a precocious entry into anaphase (19), or that it retards metaphase progression (20). Our results, showing no change, agree with neither; clearly, more work is needed on this problem.

Injection of Ca during anaphase causes interesting effects on the motion of chromosomes (43). Our results, showing that the [Ca] appeared to increase during anaphase, suggested that the level of ion might regulate the rate of chromosome motion. Accordingly, we have injected pulses of Ca or Ca-buffers into the mitotic apparatus at early-mid anaphase to transiently increase or decrease the normal cellular Ca level, and then monitored the effect on the movement of chromosomes. Control injections of potassium (Fig. 3) or chloride have no effect on the movement of chromosomes. Even though the microneedle remains stuck into the side of the mitotic apparatus the chromosomes continue normal motion. When Ca is injected, the results are quite different and depend on the level to which the ion is injected. Between the resting level (0.1 μM) and about 1 μM, Ca has no effect; however, when a

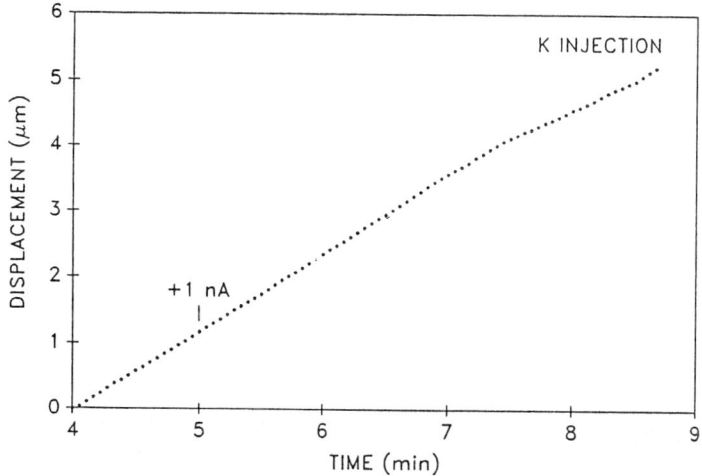

Figure 3. Ionophoretic injection of potassium ions (+1-4 nA, 10 s) into the mitotic apparatus at 5 min after the onset of anaphase has no effect on the rate of chromosome motion. Figure reproduced from Zhang et al. (43) with permission from *The Journal of Cell Biology*.

pulse is applied that reaches 1 µM (0.8-1.5 µM) the chromosomes are transiently accelerated 2-fold, from about 1 to 2 µm/s (43) (Fig. 4). At 2 µM Ca, chromosome motion is slowed, and at higher levels, motion can be briefly stopped altogether (Fig. 5). Since the Ca is applied only briefly (10 s), the cellular pumps are able to reduce the ion level (Figs. 6 and 7), thereafter chromosome motion returns to a normal, preinjection rate (Figs. 4 and 5).

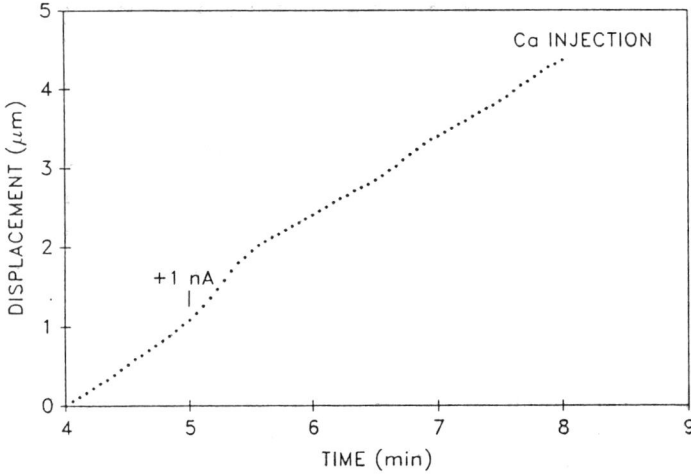

Figure 4. Injection of Ca (+1 nA, 10 s) (pipette contains 20 mM CaCl$_2$, 100 mM KCl) causes a transient increase in the rate of chromosome motion from 1.1-2.1 µm/min. Figure reproduced from Zhang et al. (43) with permission from *The Journal of Cell Biology*.

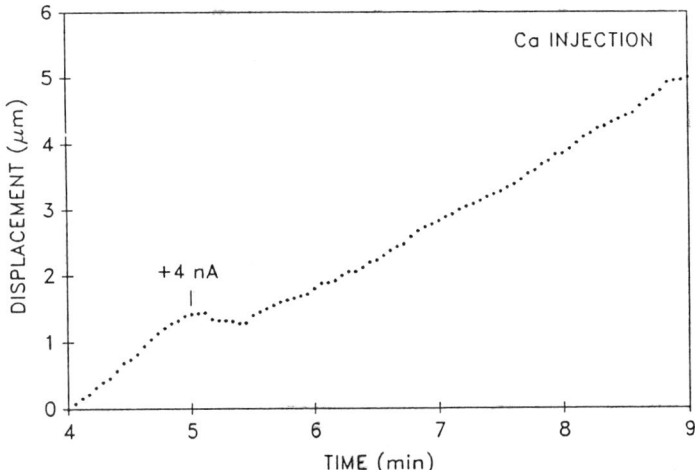

Figure 5. Injection of Ca (+4 nA, 10 s)(pipette contains 100 mM CaCl$_2$) causes brief inhibition of chromosome motion. Figure reproduced from Zhang et al. (43) with permission from *The Journal of Cell Biology*.

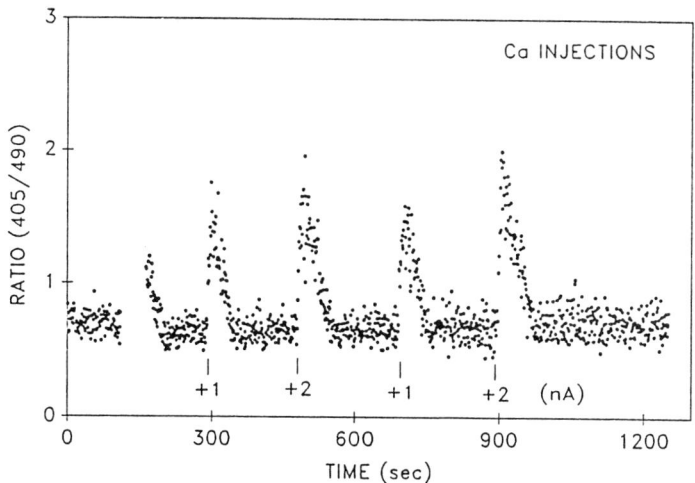

Figure 6. Ratiometric measurement of intracellular Ca in a *Tradescantia* stamen hair cell loaded with indo-1. Upward deflections of the trace at +1 and +2 correspond to the injection of the respective nA of Ca for 10 s at that moment. +1 nA of Ca stimulates chromosome motion, while +2 nA causes a slight inhibition. Figure reproduced from Zhang et al. (43) with permission from *The Journal of Cell Biology*.

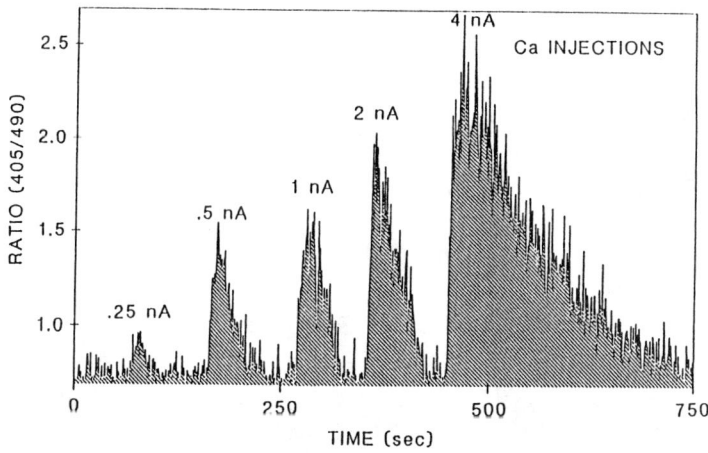

Figure 7. A comparative trace showing the relative ratios of the different levels of Ca injection. Below +1 nA, there is no effect on chromosome motion, while from +2 nA and above, there is increasing inhibition. Figure reproduced from Zhang et al. (43) with permission from *The Journal of Cell Biology*.

If the [Ca] is transiently suppressed by injection of a Ca buffer (EGTA or Br$_2$BAPTA), then chromosome motion also briefly stops (Figs. 8 and 9). In some instances, Br$_2$BAPTA ultimately causes complete anaphase arrest, possibly due to a long-lasting effect that results in the dissipation of a Ca gradient (43).

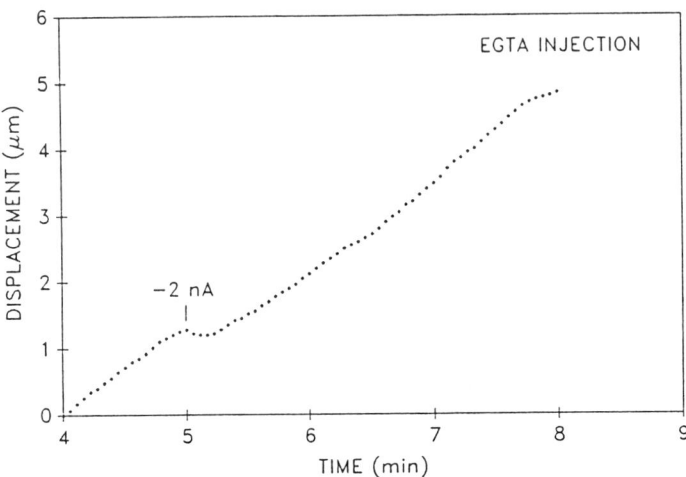

Figure 8. Injection of the Ca chelator EGTA causes a transient inhibition of chromosome motion. Figure reproduced from Zhang et al. (43) with permission from *The Journal of Cell Biology*.

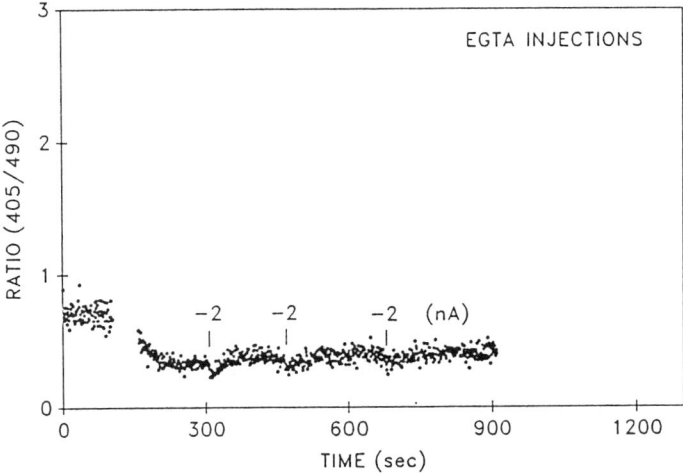

Figure 9. Injection of EGTA into an indo-1 loaded cell causes a downward deflection in the ratio. Notice also that the basal level of Ca becomes depressed. Figure reproduced from Zhang et al. (43) with permission from *The Journal of Cell Biology*.

The injections described above have all been made at the midplane of the cell, the region occupied previously by the metaphase plate and, as a consequence, both spindle halves are similarly modulated. In another series of experiments, we injected into one of the spindle pole regions, using a level of Ca (2 μM) that would cause a slowing of chromosome motion. The results,

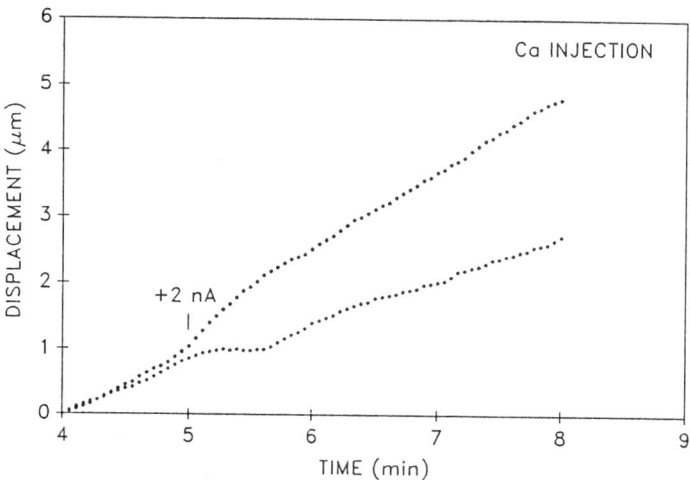

Figure 10. Injection of +2 nA of Ca into a spindle pole region inhibits chromosome motion toward that pole (lower trace) while stimulating motion to the opposite pole (upper trace). Figure reproduced from Zhang et al. (43) with permission from *The Journal of Cell Biology*.

shown in Figure 10, reveal that motion to the proximal pole (closest to the injection needle) is transiently inhibited, but that motion to the distal pole is stimulated (43). Presumably, due to diffusion from the pipette tip across the cell, the proximal pole experiences a high, inhibitory level of Ca, while the distal pole experiences a much lower [Ca] in the 1 μM range and, thus, is stimulated. These results are also interesting with regard to the coordination of the two spindle halves, since they show that motion toward the two poles can be uncoupled, whereas previous studies had indicated that motion is normally coupled (25).

These results provide strong and compelling support for a role for Ca in the control of chromosome transport. They further underscore the importance of having the [Ca] regulated to well-defined levels; thus, it is only at around 1 μM that we observe an acceleration of motion. Already, when the level reaches 2 μM, motion is inhibited. We presume that the site of Ca action is the MT; thus, it seems likely that elevated Ca facilitates depolymerization of the spindle MTs (6, 21) and permits the chromosomes to move to the poles more quickly. We are puzzled as to why the higher levels of Ca do not accelerate motion further, or at the highest level (approximately 10 μM) cause a broad-scale collapse of the mitotic apparatus. For example, studies from the Ca-MT interaction *in vitro* would suggest complete depolymerization of the MTs at these high levels (6). This does not appear to occur, as evidenced by the fact that the mitotic apparatus remains structured under these conditions and that the chromosomes do not become scrambled, rather, they simply freeze in place. Furthermore, once the high Ca has been reduced to normal levels, the chromosomes immediately resume motion to the poles, without any apparent rebuilding of the mitotic apparatus. Thus, while we are not surprised that Ca modulates the movement of chromosomes during anaphase, we are surprised by the specifics of the events as they occur within the cell and note that some of these are not predicted by what has been reported concerning the behavior of MT in the test tube under varying conditions of Ca.

Additional studies have involved the microinjection of different signaling agents, including inositol 1,4,5-trisphosphate (IP_3), GTPγS and related compounds (43). IP_3 has no effect on the rate of chromosome motion (Fig. 11) and, also, no apparent effect on the intracellular level of Ca (Fig. 12) beyond that which is caused simply by the hyperpolarization of the membrane potential resulting from the injection of any negatively charged ion. In marked contrast, GTPγS (Fig. 13) and, to a lesser extent, GTP cause a substantial increase in the rate of chromosome motion (43). For GTPγS, the stimulation of motion reaches about two-fold, as with 1 μM Ca, but may last twice as long. With GTP, the brief transient increase in motion closely resembles the effect of 1 μM Ca. These stimulatory effects are specific for GTPγS and GTP since ATPγS, ATP, and GDPβS cause no stimulation or inhibition of chromosome motion (43). We initially assumed that GTPγS and GTP might be achieving their effect through the modulation of Ca, but our results show that these agents, like IP_3, only generate a small spike similar to that caused by the injection of any negative ion, including Cl, ATPγS, ATP, and GDPβS (Fig. 14). It also seems pertinent that GTPγS, while modifying streaming in nondividing cells, does so in a manner very different from elevated Ca. The latter, when injected into the cell, quickly causes streaming to stop and with it an aggregation of cytoplasm around the pipette tip and the collapse of the

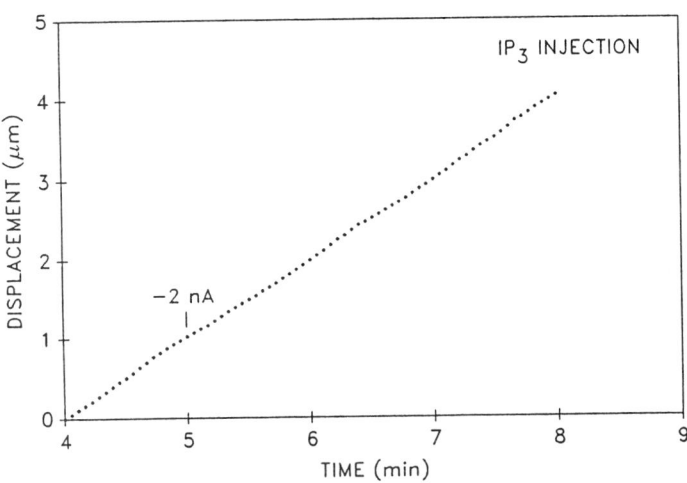

Figure 11. Injection of IP_3 has no effect on the motion of chromosomes during anaphase.

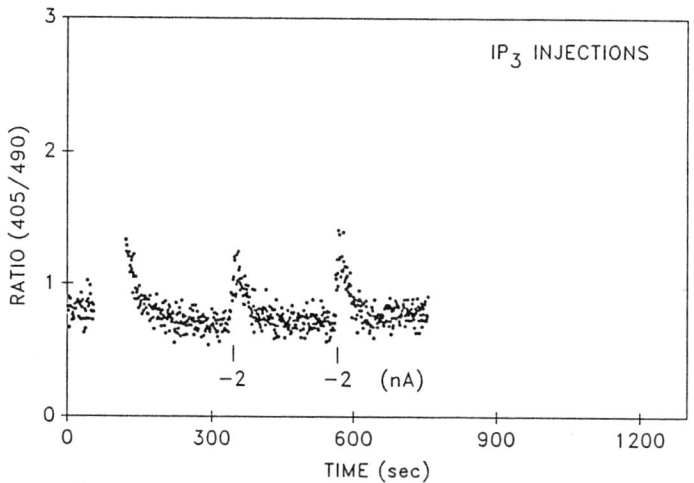

Figure 12. Injection of IP₃ generates only a small increase in the [Ca]. The ratio change corresponds to that caused by injection of between +0.25 and +0.5 nA Ca, a level that has no effect on chromosome motion.

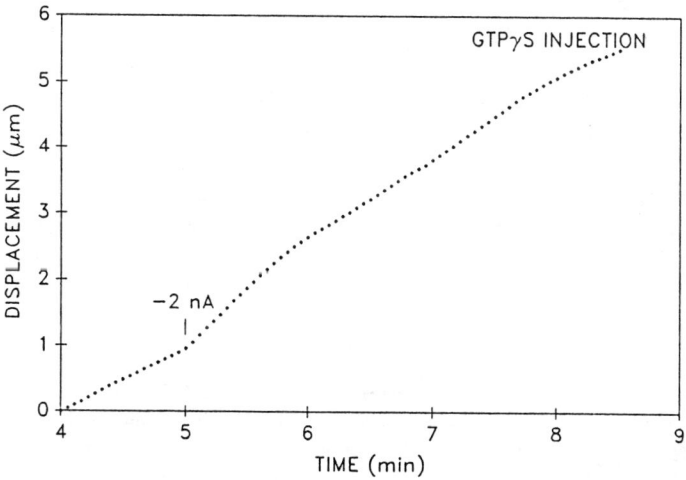

Figure 13. Injection of GTPγS (-2 nA, 10 s) stimulates two-fold for up to a min. Figure reproduced from Zhang *et al.* (43) with permission from *The Journal of Cell Biology*.

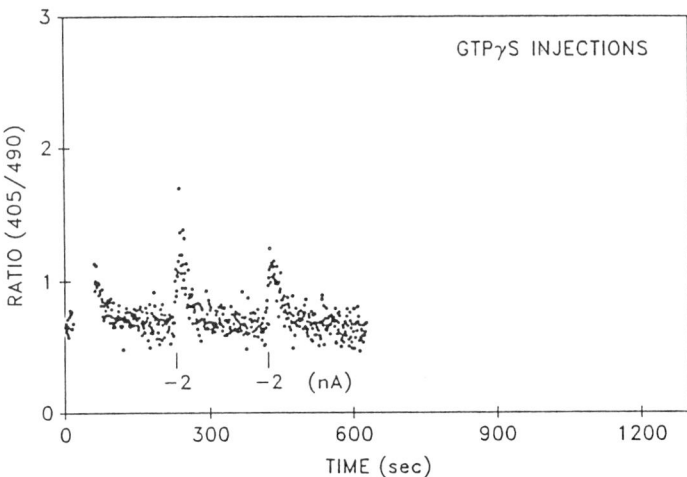

Figure 14. Injection of GTPγS (-2 nA, 10 s) generates only a small increase in the [Ca] which, by itself, would be insufficient to facilitate chromosome motion. Figure reproduced from Zhang et al. (43) with permission from *The Journal of Cell Biology*.

transvacuolar strands. GTPγS, on the other hand, causes a slowing of the large particles, but never stops streaming entirely. Its effect is also very much slower here than on anaphase chromosome motion where, by contrast, GTPγS immediately accelerates motion. Thus, these results provide evidence for a GTP-specific effect, possibly a "G" protein, that is able to stimulate chromosome motion independently from that caused by elevated Ca.

In assessing these findings, perhaps we should not be surprised that there is more than one factor that can modulate the movement of chromosomes. The mitotic apparatus, possibly more than any other motile machine, is designed for accuracy and fidelity, and not for speed. Built-in redundancies allow it to adjust to various stimuli, a situation that may relate closely to the evolutionary necessity of correctly separating the sister chromatids. In speculating on the Ca-independent mechanism of chromosome acceleration, we focus on the possible role of phosphorylation as controlled by protein kinase C. GTPγS might be stimulating phosphatidylinositol 4,5 bisphosphate hydrolysis, with the concomitant production of IP_3 and diacylglycerol (2). As pointed out above, our attempts to show a role for IP_3 produced negative results on both chromosome motion and Ca release. Although we lack experimental evidence, it nevertheless seems reasonable to suggest that diacylglycerol, which can stimulate protein kinase C without a substantial increase in [Ca] (2, 26), might be one of the key links between GTPγS and chromosome motion acceleration shown herein.

CONCLUSIONS

It seems evident that Ca does play an important role in the regulation of mitosis, although perhaps not in ways that have been predicted. Ca elevation within limited physiological levels can have a pronounced effect on chromosome motion causing a two-fold acceleration. But other factors, such as GTPγS, can also increase chromosome motion without elevating the [Ca]. The response of the mitotic apparatus to the elevated [Ca] can be partly predicted and explained as an effect of the ion on MT depolymerization; however, the results leave quite unexplained why the substantial increases of Ca do not simply destroy the mitotic apparatus. Future studies might profit greatly from a further examination of the effect of Ca on spindle elements in living cells.

The metaphase/anaphase transition is especially enigmatic regarding the role of Ca. On the one hand, it seems entirely reasonable that a pulse of Ca would trigger the splitting of the sister chromatids and initiate anaphase, but the evidence, especially from direct measurements of Ca within dividing cells, does not support that conclusion. On the other hand, many physiological studies using agents that reduce intra- and extracellular Ca or block its entry into the cell, repeatedly point to a specific requirement for this ion at the metaphase/anaphase transition. A possible way to resolve these discrepancies is to invoke a role for protein kinase-C, an enzyme that requires Ca, but not at an elevated level (2, 26). It may be that agents which reduce the [Ca] or block ion entry, profoundly alter the normal Ca milieu within the plasmalemma and, thus, destroy the activity of protein kinase-C. Within this scheme it is easier to understand how administration either of Ca or 1,2 dioctanoylglycerol to a Ca-stressed cell can allow the metaphase block to be overcome, and enable the cell to enter anaphase (22, 36). The permeant diacylglycerol analogs simply activate protein kinase C, while Ca elevation may re-establish normal conditions within the plasmalemma that facilitate kinase activity; Ca that is administered as a sudden increase may be unimportant and quite unnecessary.

Future work on the Ca-mitosis question will greatly benefit from an elucidation of the role of protein phosphorylation. In support of this assertion, I note several exciting recent developments. *i*) The nuclear envelope-associated proteins, the lamins, are phosphorylated at the time of envelope breakdown and dephosphorylated for envelope reformation (7). *ii*) Reactivation of elongation of isolated Diatom spindles requires the phosphorylation of a 205-kD protein (42). *iii*) The phosphorylation of a 62-kD protein from sea urchin mitotic apparatuses by a Ca/calmodulin-dependent kinase directly correlates with the depolymerization of spindle MTs (5). *iv*) The centrosomes of cultured rat kidney cells contain phosphorylated proteins that become dephosphorylated at the metaphase/anaphase transition (32). *v*) Finally, the discovery of mitosis-promoting factor (MPF) as a highly conserved component involved in mitotic regulation from yeasts to man further draws our attention to phosphorylation/dephosphorylation cycles (24). P34, a component of MPF, is a kinase and is, itself, subjected to cyclical phosphorylation and dephosphorylation.

Determining the possible role of Ca in regulating the kinases, and deciphering the identity and function of the protein substrates, will constitute major steps forward in our quest to elucidate the control of mitosis.

Related to the above processes is the phenomenon of Ca transients. Do they occur and, if so, when, where, and to what magnitude? The preliminary data are still too few and fragmentary to construct a complete or even correct view about them. Given the possibility that they might be highly localized, it becomes important to apply imaging and detection methods that allow one to observe Ca transients near individual kinetochore fibers. When these data are available, then we can build a realistic scheme that specifies how the normal processes of mitosis are regulated.

ACKNOWLEDGMENTS

This work has been supported by grants from the National Science Foundation (DCB 88-01750) and the United States Department of Agriculture (88-37261-3727).

LITERATURE CITED

1. Alderton JM, Kao JPY, Tsien RY, Steinhardt RA (1989) Calcium signaling during mitosis in Swiss 3T3 cells. J Cell Biol **107**: 238a
2. Berridge MJ (1984) Inositol trisphosphate and diacylglycerol as second messengers. Biochem J **220**: 345-360
3. Callaham DA, Hepler PK (1990) The measurement of intracellular free calcium in single, living plant cells. In JG McCormack, PH Cobbold, eds, Cellular Calcium - A Practical Approach. Oxford Univ Press, UK (in press)
4. Chen TL, Wolniak SM (1987) Mitotic progression in stamen hair cells of *Tradescantia* is accelerated by treatment with ruthenium red and Bay K-8644. Eur J Cell Biol **45**: 16-22
5. Dinsmore JH, Sloboda RD (1988) Calcium and calmodulin-dependent phosphorylation of a 62 kD protein induces microtubule depolymerization in sea urchin mitotic apparatuses. Cell **54**: 769-780
6. Dustin P (1984) Microtubules, 2nd Ed. Springer-Verlag, NY, p 482
7. Gerace L, Burke B (1988) Functional organization of the nuclear envelope. Annu Rev Cell Biol **4**: 335-374
8. Grynkiewicz G, Poenie M, Tsien RY (1985) A new generation of Ca^{2+}-indicators with greatly improved fluorescence properties. J Biol Chem **260**: 3440-3450
9. Harris P (1977) Triggers, trigger waves, and mitosis: a new model. In JR Jeder, ED Buetow, IL Cameron, GM Padilla, AM Zimmerman, eds, Monographs in Cell Biology. Academic Press, NY, pp 75-104
10. Henson JH, Begg DA, Beaulieu SM, Fishkind DJ, Bonder EM, Teraski M, Lebeche D, Kaminer B (1989) A calsequestrin-like protein in the endoplasmic reticulum of the sea urchin: localization and dynamics in the egg and first cell cycle embryo. J Cell Biol **109**: 149-161
11. Hepler PK (1980) Membranes in the mitotic apparatus of barley cells. J Cell Biol **86**: 490-499
12. Hepler PK (1985) Calcium restriction prolongs metaphase in dividing *Tradescantia* stamen hair cells. J Cell Biol **100**: 1363-1368

13. **Hepler PK** (1989) Calcium transients during mitosis: observations in flux. J Cell Biol **109**: 2567-2573
14. **Hepler PK** (1989) Membranes in the mitotic apparatus. *In* J Hyams, B Brinkley, eds, Mitosis: Molecules and Mechanisms. Academic Press, London, pp 241-271
15. **Hepler PK, Callaham DA** (1987) Free calcium increases during anaphase in stamen hair cells of *Tradescantia*. J Cell Biol **105**: 2137-2143
16. **Hepler PK, Wayne RO** (1985) Calcium and plant cell development. Annu Rev Plant Physiol **36**: 397-439
17. **Hepler PK, Wolniak SM** (1983) Membranous compartments and ionic transients in the mitotic apparatus. Modern Cell Biol **2**: 93-112
18. **Hepler PK, Wolniak SM** (1984) Membranes in the mitotic apparatus: their structure and function. Int Rev Cytol **90**: 169-238
19. **Izant JG** (1983) The role of calcium ions during mitosis. Calcium participates in the anaphase trigger. Chromosoma (Berl) **88**: 1-10
20. **Keith CH** (1987) Effect of microinjected calcium-calmodulin in mitosis in PtK$_2$ cells. Cell Motil Cytoskeleton **7**: 1-9
21. **Kiehart DP** (1981) Studies on the *in vivo* sensitivity of spindle microtubules to calcium ions and evidence for a vesicular calcium-sequestering system. J Cell Biol **88**: 604-617
22. **Larsen PM, Tung-Ling L Chen, Wolniak SM** (1989) Quin2-induced metaphase arrest in stamen hair cells can be reversed by 1,2-dioctanoylglycerol but not by 1,3-dioctanoylglycerol. Eur J Cell Biol **48**: 212-219
23. **McIntosh JR, Koonce MP** (1989) Mitosis. Science **246**: 622-628
24. **Murray AW, Kirschner MW** (1989) Dominoes and clocks: the union of two views of the cell cycle. Science **246**: 614-621
25. **Nicklas RB** (1979) Chromosome movement and spindle birefringence in locally heated cells: interaction versus local control. Chromosoma (Berl) **74**: 1-37
26. **Nishizuka Y** (1986) Studies and perspectives of protein kinase C. Science **233**: 305-312
27. **Poenie M, Alderton J, Tsien RY, Steinhardt RA** (1986) Calcium rises abruptly and briefly throughout the cell at the onset of anaphase. Science (Washington, DC) **233**: 886-889
28. **Ratan RR, Maxfield FR, Shelanski ML** (1988) Long-lasting and rapid calcium changes during mitosis. J Cell Biol **107**: 993-999
29. **Ratan RR, Shelanski ML, Maxfield FR** (1986) Transition from metaphase to anaphase is accompanied by local changes in the cytoplasmic free calcium in PtK$_2$ kidney epithelial cells. Proc Natl Acad Sci USA **83**: 5136-5140
30. **Steinhardt RA, Alderton J** (1988) Intracellular free calcium rise triggers nuclear envelope breakdown in the sea urchin embryo. Nature (Lond) **332**: 364-366
31. **Tombes RM, Borisy GG** (1989) Intracellular free calcium and mitosis in mammalian cells: anaphase onset is calcium modulated, but is not triggered by a brief transient. J Cell Biol **109**: 627-636
32. **Vandre DA, Borisy GG** (1989) Anaphase onset and dephosphorylation of mitotic phosphoproteins occur concomitantly. J Cell Sci **94**: 245-258
33. **Weisenberg RC** (1972) Microtubule formation *in vitro* in solutions containing low calcium concentrations. Science (Washington, DC) **177**: 1104-1105
34. **Wick SM, Hepler PK** (1980) Localization of Ca^{++}-containing antimonate precipitates during mitosis. J Cell Biol **86**: 500-513

35. Wolniak SM (1987) Lithium alters mitotic progression in stamen hair cells of *Tradescantia* in a time-dependent and reversible fashion. Eur J Cell Biol **44**: 286-293
36. Wolniak SM (1988) The regulation of mitotic spindle function. Biochem Cell Biol **66**: 490-514
37. Wolniak SM, Bart KM (1985) The buffering of calcium with quin2 reversibly forestalls anaphase onset in stamen hair cells of *Tradescantia*. Eur J Cell Biol **39**: 33-40
38. Wolniak SM, Bart KM (1985) Nifedipine reversibly arrests mitosis in stamen hair cells of *Tradescantia*. Eur J Cell Biol **39**: 273-277
39. Wolniak SM, Hepler PK, Jackson WT (1980) Detection of the membrane-calcium distribution during mitosis in *Haemanthus* endosperm with chlorotetracycline. J Cell Biol **87**: 23-32
40. Wolniak SM, Hepler PK, Jackson WT (1981) The coincident distribution of calcium-rich membranes and kinetochore fibers at metaphase in living endosperm cells of *Haemanthus*. Eur J Cell Biol **25**: 171-174
41. Wolniak SM, Hepler PK, Jackson WT (1983) Ionic changes in the mitotic apparatus at the metaphase/anaphase transition. J Cell Biol **96**: 598-605
42. Wordeman L, Cande WZ (1987) Reactivation of spindle elongation *in vitro* is correlated with the phosphorylation of a 205 kD spindle-associated protein. Cell **50**: 535-543
43. Zhang DH, Callaham DA, Hepler PK (1990) Regulation of anaphase chromosome motion in *Tradescantia* stamen hair cells by Ca and Ca-related signaling agents. J Cell Biol (in press)

TEMPORAL AND SPATIAL CHANGES IN CA^{2+} DURING PLANT DEVELOPMENT

KENNETH R. ROBINSON

*Department of Biological Sciences, Purdue University
West Lafayette, IN 47907 USA*

Abundant evidence has accumulated that an increase intracellular free calcium levels (Ca_i^{2+}) is involved in the stimulation of exocytosis in animal cells (8, 20). Examples include neurotransmitter release, cortical vesicle secretion at fertilization and insulin release from pancreatic β-cells. The Ca_i^{2+} levels required to produce a half-maximal secretory response are between 1 and 10 μM. The usual situation involves the accumulation of secretory vesicles that are induced to fuse with the plasma membrane and release their contents to the extracellular space by specific external signals that are temporally limited. These signals, often acting through G-proteins, induce a large, transient increase in Ca_i^{2+}, at least in the cortex of the target cell, which then leads to secretion. In general, all three criteria required to establish calcium's role have been met: (*i*) Ca_i^{2+} has been shown to rise by reliable measuring technique in response to the signal, (*ii*) blocking that rise blocks the secretory response, and (*iii*) raising Ca_i^{2+} in the absence of the normal signal induces secretion. It should be noted that calcium-independent exocytosis may also occur (8).

Steer (23-25) has recently reviewed the matter of calcium and its possible involvement in exocytosis in plant cells. As he points out, accumulation of secretory vesicles awaiting an appropriate exocytotic signal is rare in plants, but tip growth of various algal, fungal, and higher plant cells involves the continuous transport of vesicles to the growth site and the fusion of those vesicles with the plasma membrane there. In two cases, the rhizoid of the fucoid zygote and the pollen tube, substantial evidence, both circumstantial and direct, has been presented to support the view that Ca_i^{2+} is involved; specifically, it has been proposed that there are gradients of Ca^{2+} in the fucoid rhizoid and the pollen tube, with the concentration being higher at the growing tip. In this brief review, I will examine these data and conclusions critically and will present some new information that challenges these views.

FERTILIZATION, POLARIZATION AND GROWTH OF FUCOID ZYGOTES

There are at least three processes in the early development of *Fucus* and *Pelvetia* in which calcium has been implicated. These are fertilization, polar axis formation, and rhizoidal growth; I will discuss these separately.

Fertilization

Development is initiated in the brown algae, *Fucus* and *Pelvetia*, when the gametes are released into the intertidal sea and the motile sperm cell fertilizes the egg. It has become clear that fertilization in these lower plants is similar in many ways to fertilization of animal eggs, where the role of calcium is well established (7). When the eggs of both *F. vesiculosus* and *P. fastigiata* are fertilized, a transient depolarization of the membrane potential occurs (18). This depolarization is partly dependent on external sodium and when fertilization is carried out in low-sodium sea water, increased rates of polyspermy are observed (1). Likewise, it has been observed that *F. vesiculosus* eggs can be "partially activated" by calcium ionophores; this process involves the secretion of cell wall materials and the later formation of a rhiziod-like protuberance (2). These results suggest that fucoid egg activation shares many similarities with animal egg activation. Given the central role of Ca_i^{2+} increases in animal eggs, it might be expected that there are increases in Ca_i^{2+} during fertilization of *Fucus* and *Pelvetia* eggs. There is preliminary evidence that this is so. Figure 1 shows the response to insemination of 12 *F. vesiculosus* eggs that were microinjected with aequorin (SH Brawley and KR Robinson, unpublished). Before adding sperm, about 1 counts per second above background was detectable by the photomultiplier. Within 5 min of insemination, a large increase in light was detected. This increase was sustained over the ensuing 20 min, and there appeared to be multiple peaks. Because the resting level of light emission was not measurable for a single egg, we cannot say how much it increased, but at the peaks, the light emission was at least 100 times greater than resting levels, indicating a Ca_i^{2+} level 10 times greater than the basal level if we assume that a peak is composed mostly of light from a single egg. It will be important to repeat these experiments using single eggs, improved aequorins and better detectors; nevertheless, these data strongly suggest that a large increase in Ca^{2+} occurs when fucoid eggs are fertilized and this increase triggers the initial secretion of cell wall components.

Polar axis formation

One reason for the continued fascination with *Fucus* and *Pelvetia* development is the fact that the eggs are unpatterned at the time of fertilization. The zygotes then develop a primary axis in response to external directional cues such as unilateral light and express that choice by producing a rhizoidal bulge. First cell division is then highly unequal, separating a prorhizoidal cell from a prothallus. The question then is how does the zygote

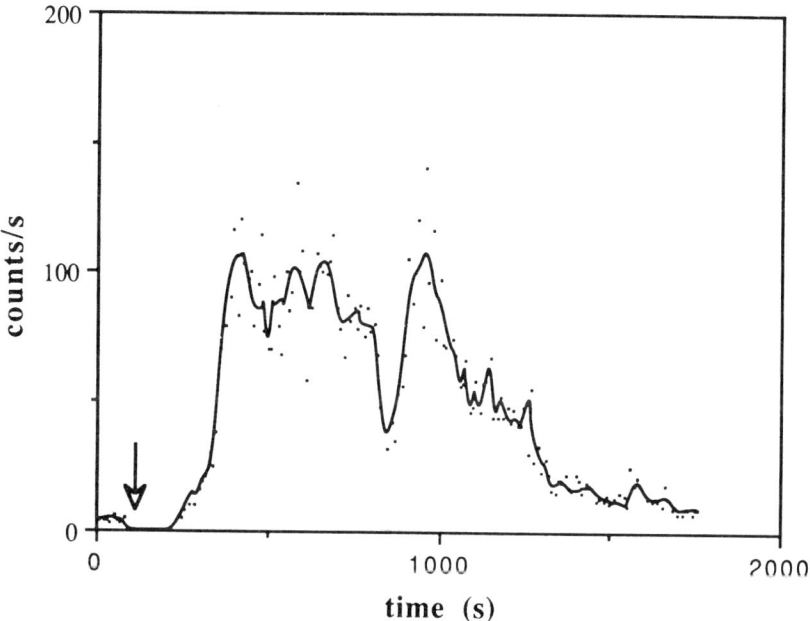

Figure 1. Aequorin was injected into 12 *Fucus vesiculosus* eggs. The eggs were in a chamber that could be placed within 1 mm of the photocathode of a low-noise photomultiplier tube. The points shown indicate the number of photons per second, averaged over 10-s intervals, recorded by the photon-counting equipment. At the time marked by the arrow, the chamber was removed, a sperm suspension added, and the chamber returned to the measuring position. A large increase (at least 100-fold) in the rate of light emission is apparent and there are multiple peaks. The light emission then approached, but did not reach, the initial background levels.

convert the external cue, which acts at the cell surface, into an intracellular response that leads to growth at one, and only one, locus. Four lines of evidence have lent support to the notion that an intracellular gradient of Ca^{2+} is a critical part of the feedback mechanism involved in polarity. First is the finding that an externally-detectable ionic current enters the future rhizoidal pole of the ungerminated zygote, and a fraction of that entering current seems to be Ca^{2+} (12). Second, there is marked asymmetry of the permeability of the plasma membrane of the zygote to Ca^{2+}. When exposed to unilateral light, the flux of Ca^{2+} into the cell, as measured by ^{45}Ca, is about five times greater on the shaded side than on the illuminated side; it is the shaded side that becomes the rhizoid (19). This asymmetry in flux is largest a few hours before overt germination and declines with the approach of germination. Both the electrical measurements and the tracer flux measurements show that calcium enters the future rhizoid and imply the existence of a cytoplasmic Ca^{2+} gradient, with higher calcium on the rhizoidal side.

The third piece of evidence for the involvement of calcium gradients is the finding that *Pelvetia* zygotes, when grown in the dark in a gradient of calcium ionophore, show a strong tendency to form their rhizoids on the sides that are exposed to the high concentration of ionophore (17). Presumably, that side of the cell is made more permeable to Ca^{2+}, and an intracellular Ca^{2+} gradient is imposed on the cell.

Finally, it has been shown recently that the injection of buffers with dissociation constants (K_D's) for Ca^{2+} in the physiological range can block germination (22). The optimal K_D was about 5 μM and buffers with K_D's between 0.4 μM and 98 μM were effective to some degree. Speksnijder *et al.* conclude that Ca^{2+} at the growth point must be about 10 μM.

Lacking are any direct measurements of the Ca^{2+} distribution in fucoid zygotes; until these are obtained, any conclusions must be tentative. Also, there are two sets of experiments that argue against the existence or involvement of Ca^{2+} gradients in the polarization of these cells. Kropf and Quatrano (9) have stained ungerminated *Fucus distichus* zygotes with chlorotetracycline (CTC), a fluorescent probe for membrane-associated calcium, and they found "no asymmetries in the distribution of membrane-associated Ca^{2+}." They also found that the removal and chelation of extracellular calcium had little effect on light-induced polarization. The value of CTC as a measure of Ca_i^{2+} distribution will be discussed below.

Rhizoidal growth

The case for the existence of gradients of Ca^{2+} in the growing rhizoid of *Fucus* embryos is more direct. Brownlee and Wood (4) have used calcium-sensitive microelectrodes to demonstrate a gradient of Ca^{2+} in the rhizoidal tip of *Fucus serratus*. They report a value for Ca_i^{2+} of 2500 nM at the tip and 300 nM in the sub-tip region 20 to 30 μM behind the tip. This gradient rapidly collapsed when Ca^{2+} was removed from the medium or La^{3+} was added. More recently, Brownlee and Pulsford (3) have microinjected the fluorescent Ca indicator dye, {1-[2-(5-carboxyoxazol-2yl)-6-aminobenzofuran-5-oxy]-2-(2-amino-5'-methylphenoxy)-ethane-N,N,N',N'-tetraacetic acid} (Fura-2), into the apical cells of *F. serratus* rhizoids. Fura-2 exhibits a shift in its excitation spectrum upon binding calcium; thus, if two appropriate excitation wavelengths are used, the ratio of the emissions gives a measure of Ca^{2+} that is independent of the dye concentration. This property makes fura-2 the ideal choice for the detection of spatial gradients of Ca^{2+}. In about half of the cases that they studied, a tip gradient was seen with Ca^{2+} at the apex being 450 nM and about 100 nM in the subapical regions. In other cases, no gradients were seen.

Kropf and Quatrano (9) also reported a calcium gradient in the rhizoids of *F. distichus*, as determined by CTC fluorescence. They found that the fluorescence was localized to the entire rhizoidal protuberance in the germinated zygote, and to a substantial portion of the apical cell in older embryos.

While all three of the methods employed in these experiments—CTC, calcium microelectrodes, and fura-2—seem to show a gradient of cytoplasmic Ca^{2+}, at least in some cases, it would be a mistake, I think, to conclude that they are mutually reinforcing. The fura-2 measurements of Brownlee and Pulsford (3) are in direct contradiction to the earlier microelectrode measurements of Brownlee and Wood (4), at least with respect to both basal and apical magnitudes of Ca_i^{2+}. Furthermore, no gradients at all could be detected by fura-2, in some cases, which raises the question of the necessity of gradients. Brownlee and Pulsford (3) point out the Ca_i^{2+} level that they measured at the tip of rhizoids that had a gradient probably wasn't high enough to promote exocytosis, which generally requires 1 to 10 μM Ca^{2+}. While the CTC determinations cannot be quantitated with respect to Ca^{2+} levels, the observed pattern of fluorescence did not coincide with the Ca^{2+} distribution seen with fura-2. Finally, it should be noted that the calcium influx asymmetry between rhizoid and thallus was maximal during photopolarization and declined to nearly zero in germinating zygotes and older embryos (16, 19). In conclusion, there is no compelling evidence that gradients of Ca_i^{2+} are involved in any necessary way in tip growth of fucoid rhizoids. While there is a general requirement for calcium for growth, reducing the calcium in the external medium to 1% of normal causes only a 50% decrease in the growth rate of *Fucus distichus* rhizoids (9).

POLLEN TUBE GROWTH

The pollen tube of higher plants is an extreme example of the phenomenon of tip growth. The pollen tube may be many millimeters long and may grow at rates as fast as 10 $\mu m/s$ (10). Calcium has long been implicated in the control of the massive secretion that occurs in distal few μm of tube, and evidence has been presented to support the idea that proximal-distal gradients of Ca_i^{2+} are a necessary component of polarized growth. Here, I will review that evidence and will present some new data that bears out this idea.

Weisenseel *et al.* (27) and Weisenseel and Jaffe (26) studied the pattern of ionic current associated with the growth of the lily pollen tube. These currents enter most of the length of the tube and leave the grain, with the maximum inward current density about midway between the tube tip and its base. While calcium was not shown to be a component of the inward current, it was argued that calcium controlled the entry of the dominant inward-current carrying ion, potassium. L. A. Jaffe *et al.* (6) exposed growing lily pollen to radioactive calcium and, after washing, froze the pollen in liquid nitrogen and pressed it against a β-sensitive emulsion. The emulsion was then exposed, at -78°C, for several days before developing. They found that the grain density near the tip was nearly always higher than at more proximal regions of the tube, regardless of whether the time of incubation in ^{45}Ca was 1 min or 5 h. In some cases, intense accumulation of grains was seen at the tip. These results suggest that calcium enters the growing pollen tube's tip at a greater rate than elsewhere and may create gradients of Ca^{2+}.

Direct evidence for the existence of calcium gradients in lily pollen tubes has been presented by Reiss, Herth, and colleagues. They reported a gradient of CTC fluorescence and a gradient of total calcium, with higher calcium at the tip (14). Other work by these authors has shown that both calcium ionophore and the organic calcium channel blocker, nifedipine, perturb pollen tube growth (5, 13). More recently, Reiss and Nobiling (15) and Nobiling and Reiss (11) loaded 2-((2-bis-(carboxymethyl)amino-5-methylphenoxy)methyl)-6-methoxy-8-bis(carboxymethyl)aminoquinoline (Quin-2), via the Acetoxy Methylester (AM), into lily pollen tubes and observed a gradient of fluorescence, again with higher fluorescence at the tip, although the gradient was less steep than the one detected with CTC.

This accumulated evidence has led to general acceptance of the idea that there are gradients of Ca_i^{2+} in growing pollen tubes and that these gradients are essential control elements of the growth process (23, 24). Recent work in my laboratory addresses the requirement for Ca^{2+} gradients in lily pollen growth. Keerti Rathore and John Cork in this laboratory have injected indo 1 into tubes of lily pollen by iontophoresis and have visualized Ca^{2+} by means of a video image processing system. We have now analyzed about 10 growing pollen tubes. In about half of these cases, a small gradient was observed in tubes that were growing during and after the measurement. In these tubes, Ca^{2+} in the distal 10 μm was no more 50% higher than the basal levels in other regions of the tube (Fig. 2). In the other cases, a definite gradient existed with Ca^{2+} in the distal 10 μm being more than twice basal levels; in some cases, this was in tubes that temporarily stopped growing in response to the UV excitation light, swelled at the tip and then resumed growing. The gradients formed during the swelling and persisted after growth resumed; in these tubes, Ca^{2+} at the tip was as much as five times as large as basal levels. We conclude preliminarily that steep gradients of Ca^{2+} are not a necessary aspect of pollen growth, since we see vigorously growing tubes that have only shallow gradients. We also note that high Ca^{2+} at the tip is associated in some cases with damage and interrupted growth and suggest that Ca^{2+} measurements must be correlated carefully with growth. Pollen tubes filled with UV-absorbing dyes are sensitive to the intense UV irradiation necessary for video imaging. A remaining unanswered question is whether or not steep gradients might exist in a very restricted region (1 μm or less) just inside the plasma membrane. Our methods would not detect such small regions of high calcium, yet they might be functionally important. One approach might be to use aequorin. Its luminescence increases with the square of the Ca^{2+} concentration, so it might be possible to visualize restricted regions of high calcium, using an imaging photon detector.

What of the above-cited evidence for Ca^{2+} gradients? None of it is sufficiently definitive to be in serious conflict with our measurements. CTC fluorescence has not been established as a reliable monitor of cytosolic Ca^{2+}.

The observed gradients may represent differences in amount of CTC in the plasma membrane. Also, some plasma membrane-associated CTC may have access to extracellular calcium. The use of the cell-permeant form of Quin-2 (11) is also questionable. We have found that loading pollen tubes by bathing

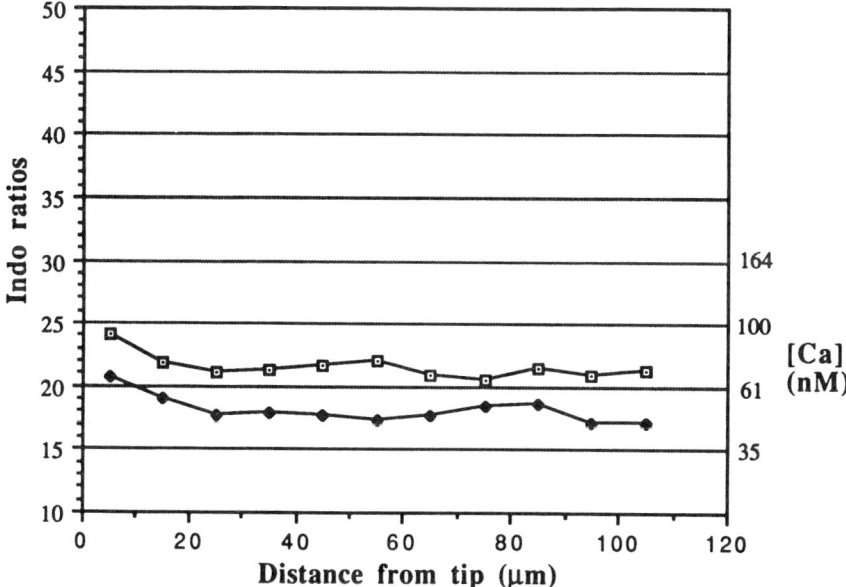

Figure 2. The Ca^{2+} distribution in two indo 1-injected lily pollen tubes is shown. Processed images were sampled in 10 μM-diameter spots centered at the indicated points. Both tubes were growing before and after the measurements. The absolute Ca^{2+} levels shown on the right hand axis are somewhat uncertain; better calibration procedures are being developed.

them in indo 1-AM gives a large apparent Ca^{2+} gradient that is not seen when the free acid is microinjected into the tube. A similar phenomenon has been reported in neurites (21) when the calcium distribution in cells loaded via the AM form of fura-2 were compared with those that were microinjected. They attributed the gradients seen in the AM-loaded neurites to the accumulation of the indicator in vesicles. We have seen similar apparent gradients when we loaded pollen tubes with indo 1-AM; thus, we suspect that the gradients reported by Nobiling and Reiss (11) are artifactual. Finally, it is hardly surprising that chemicals such as A23187 and nifedipine perturb growth. Overall calcium regulation is clearly important to cells.

In summary, the issue of the role of calcium gradients in plant tip growth is unsettled. The levels of Ca^{2+} that occur in an activating egg approach 1 μM; in pollen tubes we rarely see levels exceeding 100 nM. Likewise, in the *Fucus* rhizoid, no gradients were seen in half of the cells that were studied.

We are continuing our study of this important question. In addition, we plan to measure the distribution of H^+ in lily pollen tubes.

ACKNOWLEDGMENTS

I thank my collaborators Keerti Rathore, John Cork, and Susan Brawley for permisssion to use unpublished data. Research in my laboratory was supported by grants from the National Science Foundation and the U.S. Department of Agriculture.

LITERATURE CITED

1. **Brawley SH** (1987) A sodium-dependent, fast block to polyspermy occurs in eggs of fucoid algae. Dev Biol **124**: 390-397
2. **Brawley SH, Bell E** (1987) Partial activation of *Fucus* eggs with calcium ionophores and low-sodium seawater. Dev Biol **122**: 217-226
3. **Brownlee C, Pulsford AL** (1988) Visualization of the cytoplasmic Ca^{2+} gradient in *Fucus serratus* rhizoids: Correlation with cell ultrastructure and polarity. J Cell Sci **91**: 249-256
4. **Brownlee C, Wood JW** (1986) A gradient of cytoplasmic free calcium in growing rhizoid cells of *Fucus serratus*. Nature **320**: 624-626
5. **Herth W** (1978) Ionophore A23187 stops tip growth, but not cytoplasmic streaming, in pollen tubes of *Lilium longiflorum*. Protoplasma **96**: 275-282
6. **Jaffe LA, Weisenseel MH, Jaffe LF** (1975) Calcium accumulations within the growing tips of pollen tubes. J Cell Biol **67**: 488-492
7. **Jaffe LF** (1983) Sources of calcium in egg activation: A review and hypothesis. Dev Biol **99**: 265-276
8. **Knight DE, von Grafenstein H, Athayde CM** (1989) Calcium-dependent and calcium-independent exocytosis. Trends Neurosci **12**: 451-458
9. **Kropf DL, Quatrano RS** (1987) Localization of membrane-associated calcium during development of fucoid algae using chlorotetracyline. Planta **171**: 158-170
10. **Mascarenhas JP** (1989) The male gametophyte of flowering plants. The Plant Cell **1**: 657-664
11. **Nobiling R, Reiss H-D** (1987) Quantitative analysis of calcium gradients and activity in growing pollen tubes of *Lilium longiflorum*. Protoplasma **139**: 20-24
12. **Nuccitelli R** (1978) Ooplasmic segregation and secretion in the *Pelvetia* egg is accompanied by a membrane-generated electrical current. Dev Biol **62**: 13-33
13. **Reiss H-D, Herth W** (1985) Nifedipine-sensitive calcium channels are involved in polar growth of lily pollen tubes. J Cell Sci **76**: 247-254
14. **Reiss H-D, Herth W, Nobiling R** (1985) Development of membrane- and calcium-gradients during pollen germination of *Lilium longiflorum*. Planta **163**: 84-90
15. **Reiss H-D, Nobiling R** (1986) Quin-2 fluorescence in lily pollen tubes: Distribution of free cytoplasmic calcium. Protoplasma **131**: 244-246
16. **Robinson KR** (1973) Ion movements and membrane asymmetries in the developing fucoid egg. PhD Thesis, Purdue University
17. **Robinson KR, Cone R** (1980 Polarization of fucoid eggs by a calcium ionophore gradient. Science **207**: 77-78
18. **Robinson KR, Jaffe LA, Brawley SH** (1981) Electrophysiological properties of fucoid algal eggs during fertilization. J Cell Biol **91**: 179a

19. **Robinson KR, Jaffe LF** (1975) Polarizing fucoid eggs drive a calcium current through themselves. Science **187**: 70-72
20. **Rubin RP** (1982) Calcium and Cellular Secretion. Plenum Press NY
21. **Silver RA, Lamb AG, Bolsover SR** (1989) Elevated cytosolic calcium in the growth cone inhibits neurite elongation in neuroblastoma cells: Correlation of behavioral states with cytosolic calcium concentration. J Neurosci **9**: 4007-4020
22. **Speksnijder JE, Miller AL, Weisenseel MH, Chen T-H, Jaffe LF** (1989) Calcium buffer injections block fucoid egg development by facilitating calcium diffusion. Proc Natl Acad Sci USA **86**: 6607-6611
23. **Steer MW** (1988) The role of calcium in exocytosis and endocytosis in plant cells. Physiol Plant **72**: 213-220
24. **Steer MW** (1989) Calcium control of pollen tube tip growth. Biol Bull **176**(S): 18-20
25. **Steer MW, Steer JM** (1989) Tansley review No 16 Pollen tube tip growth. New Phytol **111**: 323-358
26. **Weisenseel MH, Jaffe LF** (1976) The major growth current through lily pollen tubes enters as K^+ and leaves as H^+. Planta **133**: 1-7
27. **Weisenseel MH, Nuccitelli R, Jaffe LF** (1975) Large electrical currents traverse growing pollen tubes. J Cell Biol **66**: 556-567

Calcium in Plant Growth and Development, Robert T. Leonard and Peter K. Hepler, Eds, 1990, The American Society of Plant Physiologists Symposium Series, Vol. 4

CALCIUM ION CURRENTS AND GRADIENTS IN FUCOID EGGS

Lionel F. Jaffe

Marine Biological Laboratory, Woods Hole, MA 02543, USA

This paper briefly discusses the older literature on calcium and localization in fucoid eggs as well as two recent papers. One of these pursues this problem by using BAPTA-type calcium buffer injections (or "baptism"); the other by using a newly developed, vibrating calcium electrode. It also presents some preliminary new findings on the use of hyperosmotic media to delay or even reverse localization of a rhizoidal outgrowth.

Development of the fucoid egg is a prototype of the localization problem in plant cells (7). Unlike all animal eggs, the unfertilized fucoid egg is practically apolar. Localization, polarization, symmetry-breaking--call it what you will--this first step in pattern formation can be studied from the beginning.

The most convincing single evidence for its apolar nature remains the fact that large numbers of twin forms, with rhizoids formed at two opposite poles, develop in response to illumination with weak, bilateral polarized light (5). This unique stimulus affects two opposite poles in each cell to exactly equal degrees and thus generates twins. However, numerous other stimuli can localize rhizoidal outgrowth at a single pole. The important natural localizers are unilateral unpolarized light and some remarkable, but still unidentified, diffusible substance emitted by all marine plants. Other natural localizers are flow, nearby eggs, and the point of sperm entry. Unnatural, but potent, localizers include electrical fields and imposed gradients of osmotic strength, pH, dinitrophenol, K^+, and calcium ionophores.

All of these vectors must act through an inner loop which amplifies almost any small difference. I have put forward the hypothesis that this inner loop includes a localized concentration of calcium channels which raises the concentration of free calcium in the region under these channels. The older evidence for this hypothesis included direct $^{45}Ca^{++}$ evidence of an early calcium current through polarizing fucoid eggs; congruent evidence of steady early electrical currents through such eggs; evidence of localized early secretion; as well as the formation of rhizoids toward a source of calcium ionophore. Recently, however, Kropf and Quatrano have challenged the calcium theory of localization (9).

One of their arguments is based upon their finding that fucoid eggs can be polarized by unilateral light applied in the complete absence of external calcium. However, there are two plausible alternative explanations available: Calcium may well enter the former dark pole after such eggs are returned to 10 mM Ca^{++} seawater and before they grow out; sodium may flow through localized calcium channels to locally release calcium from internal stores during low-calcium treatment. Their other argument is based upon their chlorotetracycline (CTC) images of a vivid, bright line (or really sheet) just under the growing tip of fucoid eggs as they begin to grow out. They suggest that a failure to find comparable localizations of CTC light before outgrowth argues against the calcium leak theory. However, their methods may well have lacked the sensitivity needed to see earlier, but fainter, CTC images. Furthermore, as is reported below, we now have evidence that polarization is still reversible at the early outgrowth stage when localized CTC light was so clearly seen.

BAPTISM

Nevertheless, in response to Kropf and Quatrano's challenge, we have carried out a new test of the calcium leak hypothesis using injections of various calcium buffers (14). Our approach, which we call "baptism" after the BAPTA-type buffers used, was based on the concept of facilitated diffusion. Suppose that growth localization really requires the maintenance of a zone of high calcium under the future growth pole. Maintenance of such a zone would surely require a steady localized influx of calcium together with steady uptake by some inner membrane, perhaps that of the ER, or of secretory vesicles. Moreover, it is possible that the needed flow between the source and sink membranes would be diffusion limited so as to set up a steady calcium gradient. Under these circumstances, introduction of enough calcium buffer to speed diffusion (by shuttling the calcium) would reduce the level of subsurface calcium supposedly needed for local outgrowth. In fact, appropriate buffer injections regularly and permanently block outgrowth without visibly damaging the eggs. Moreover, when the efficacy of a series of BAPTA-type buffers of different dissociation constants (or K_D's) was compared, a striking peak in efficacy was found for buffers with a K_D of about 5 μM. Both weaker and stronger buffers were less effective. Moreover, they were less effective to just the degree predicted by facilitated diffusion theory if a zone of high calcium of the order of 5 to 10 μM must be maintained somewhere to support localization of an outgrowth. It might be argued—as Roger Tsien has pointed out—that the needed calcium gradient is somewhere inside of an extended organelle such as the ER. However, what is known of free calcium levels within the ER suggests that they are far too high to be optimally shuttled by buffers with K_D's of about 5 μM (4, 11). This indicates that the needed high calcium zone is somewhere in the cytosol and presumably right under the plasma membrane. The "baptism" results, then, are further good evidence of the need for such a zone to support local outgrowth.

THE VIBRATING CALCIUM ELECTRODE

Ionic currents before outgrowth have been studied using $^{45}Ca^{++}$ and using a vibrating extracellular voltage-sensitive electrode (8). While these methods are quite reliable, they have serious limitations. $^{45}Ca^{++}$ has been used with large populations of eggs in parallel and, used this way, provides absolutely minimal spatial and temporal resolution; while the vibrating voltage-sensitive electrode maps and measures net electrical current rather than individual ionic currents such as calcium ion currents. In an effort to overcome these limitations, we have now developed a calcium-specific vibrating electrode (10). The principle of this device is to measure minute standing gradients of free calcium outside of local sinks and sources of calcium at a cell's surface. Standing differences as small as 0.01% can be detected, and differences as small as 0.1% measured between points 4 to 40 μm apart. In exploratory experiments, we have already measured steady calcium currents entering the tips of growing *Pelvetia* (fucoid) rhizoids and growing tobacco pollen. In due course, it may prove possible to measure such currents into the nascent outgrowths before they form.

EFFECTS OF HYPOOSMOTIC MEDIA ON LOCALIZATION IN FUCOID EGGS

The older literature contains two very interesting indicators of such a role. First, *Pelvetia* eggs can be strongly polarized by exposure to steady osmotic gradients as small as 5% with outgrowths induced at the hypotonic end (1). Second, large pulses of electrical current can be induced to enter growing *Pelvetia* rhizoids by reductions in the osmolarity of the seawater as small as 5%. Moreover, an analysis of this phenomenon showed that these electrical pulses are part of a pressure-relief mechanism induced by local calcium ion entry (12). Presumably, some stretch-sensitive calcium channels are activated by small increases in cell turgor pressure. The resultant entry of calcium then probably induces the transient fusion of the new organelle recently discovered by Gilkey and Staehelin in fucoid eggs (3). The chloride channels in the membrane of these organelles then provides for local chloride efflux. Finally, this, plus the resultant global potassium efflux together, provide pressure relief.

All of this suggests the hypothesis that the gradual rise in turgor pressure before fucoid eggs germinate helps trigger the increased transcellular electrical current which is known to accompany germination or initial outgrowth (6). It also suggests that this rise in net current reflects a natural rise in local calcium influx during germination, which is comparable to those induced by sudden increases in turgor pressure. It also raises the question as to what extent hyperosmotic media may act to delay or even reverse the localization of tip growth. The last part of this brief paper will present a preliminary account of some new experiments on such effects.

EFFECTS OF HYPEROSMOTIC MEDIA ON THE LOCALIZATION OF TIP GROWTH

It has long been known that if fucoid eggs are grown in media made sufficiently hyperosmotic with various sugars (say, about 0.5 M or more), they will undergo numerous cell divisions without forming an outgrowth (15). I can now report that such media also greatly delay localization. Figure 1 graphs data from one of three similar experiments which establish this point. Normally, the directions of rhizoid initiation in an egg population are completely insensitive to light after about 18 h. Yet, even in 0.36 to 0.45 M

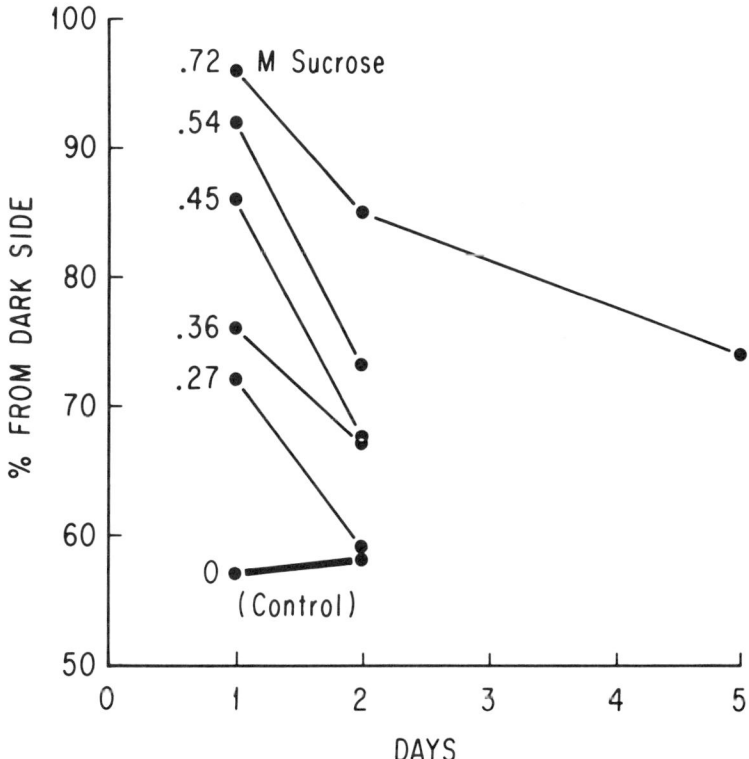

Figure 1. Hyperosmotic media can greatly delay localization in *Pelvetia* (fucoid) eggs. Four hours after fertilization, eggs were shifted from natural seawater to natural seawater made hyperosmotic with the indicated molarity of sucrose. They were cultured in darkness for 1, 2, or 5 d and then shifted back to natural seawater under unilateral white light from a nearby fluorescent lamp. A few days later, 100 eggs in each dish were scored for outgrowths which began toward or away from the light. The small deviation from random (or 50%) in the controls probably results from the fact that outgrowths beginning toward the light then bend toward the dark so as to somewhat obscure the point of origin. T = 14°C (experiment of 6/6/89).

sucrose--which scarcely plasmolyzes the eggs--there is a substantial effect of unilateral light applied as late as 2 d; moreover, in 0.72 M sucrose, substantial sensitivity remains even after 5 d. Furthermore, the slow decline in sensitivity seems to roughly, but not exactly, parallel visible germination in the hyperosmotic media (data not shown). Apparently, "commitment" to outgrowth localization can be so easily delayed that it cannot be usefully thought of as being controlled by an internal clock.

A more surprising new finding is that mild turgor reduction, combined with light reversal, efficiently reverses polarity after outgrowths form (at 13 h). In other words, after such treatment, the first outgrowth aborts and a second antipodal one forms and generates a rhizoid (Fig. 2A). On the other hand, in controls subjected to light reversal without turgor reduction, the original outgrowth does not abort; instead, it curves so as to reverse direction (Fig. 2B). Moreover, this same "U-turn" develops if turgor reduction and light reversal is postponed to 19 h (Fig. 2C).

Table I provides quantitative data from my one extensive experiment of this type. It will be seen that at 12.7 h--when 78% of the eggs had visibly germinated and some, no doubt, invisibly germinated--that practically all of the germinated eggs are induced to form an antipodal outgrowth by light reversal in 0.6 M sucrose seawater (which visibly plasmolyzes); while about 70% are so induced in 0.4 M sucrose (which does not visibly plasmolyze). However, by 19.2 h, no more than 5% of the eggs form second, antipodal outgrowths (and even these may be artifacts resulting from egg movement during media changes).

Evidently, the commitment to local tip growth remains easily reversible for some hours after an outgrowth forms. One can imagine two mechanisms for such osmotic depolarization: turgor reduction may close stretch-sensitive calcium channels (2); while plasmolysis may disrupt local growth by separating the plasma membrane from cellulose-generating rosettes in the wall (13). Moreover, these two mechanisms could well act together to generate these remarkable results.

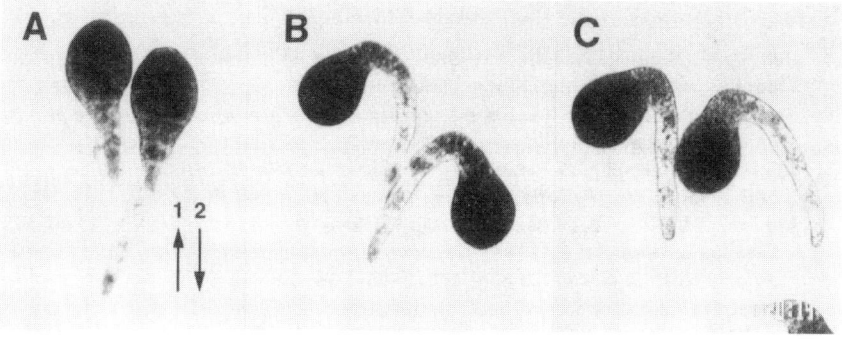

Figure 2. Typical responses of recently germinated *Pelvetia* eggs to turgor reduction together with light reversal. A, Eggs illuminated in the direction of the first arrow from the time of fertilization until 12.7 h later in natural seawater; then shifted to seawater plus 0.4 M sucrose with light direction reversed, as shown by second arrow; finally shifted back to natural seawater (with light still in second direction) at 40 h and photographed at 5-1/2 d. B, control eggs without hyperosmotic treatment. C, Eggs transferred to 0.4 M sucrose with light reversal, etc., at 19 h. Remnants of the aborted rhizoidal outgrowths away from the first light (in Fig. 2A) are not easily distinguished from initial apical papillae in (Fig. 2B and 2C); however, these first outgrowths were unmistakeably visible at 13 h in 78% of the eggs. T = 14.5°C (experiment of 7/12/90).

Table I. *Main Outgrowth Initiation Directions in Eggs Photoreversed and Transferred to More-or-Less Hyperosmotic Seawater for a Day (as described and illustrated in Figure 2)*

Directions were observed at 5 d. Each figure is the percentage growing away from the <u>second</u> light as obtained by counting 100 eggs.

	Reversal Time	
Sucrose (M)	12.7 h[a]	19.2 h[b]
0	15	3
0.4	79	7
0.6	99	4
1.0	94	2

[a]78% visibly germinated at this time.
[b]100% visibly germinated at this time.

ACKNOWLEDGMENT

The new experiments described here were supported by a competitive research grant from the USDA #8801220.

LITERATURE CITED

1. **Bentrup F, Sandan T, Jaffe L** (1967) Induction of polarity in *Fucus* eggs by potassium ion gradients. Protoplasma **64**: 254-266
2. **Falke LC, Edwards KL, Pickard BG, Misler E** (1988) A stretch-activated anion channel in tobacco protoplasts. FEBS Lett **237**: 141-144
3. **Gilkey JC, Staehelin LA** (1989) A new organelle related to osmoregulation in ultrarapidly frozen *Pelvetia* embryos. Planta **178**: 425-435
4. **Henson JH, Begg DA, Beaulieu SM, Fishkind DJ, Bonder EM, Teresaki M, Lebeche D, Kaminer B** (1989) A calsequestrin-like protein in the endoplasmic reticulum of the sea urchin egg. J Cell Biol **109**: 149-161
5. **Jaffe L** (1956) Effect of polarized light on polarity of *Fucus*. Science **123**: 1081-1082
6. **Jaffe L** (1966) Electrical currents through the developing *Fucus* egg. Proc Natl Acad Sci USA **56**: 1102-1109
7. **Jaffe L** (1968) Localization in the developing *Fucus* egg and the general role of localizing currents. Adv Morphogenesis **7**: 295-328
8. **Jaffe LF** (1982) Developmental currents, voltages and gradients. Symp Soc Dev Biol **40**: 183-218
9. **Kropf DL, Quatrano RS** (1987) Localization of membrane-associated calcium during development of fucoid algae using chlorotetracycline. Planta **171**: 158-170
10. **Kuhtreiber WM, Jaffe LF** (1990) Detection of extracellular calcium gradients with a calcium specific vibrating electrode. J Cell Biol (in press)
11. **MacLennan DH, Campbell KP, Reithmeier RAF** (1983) Calsequestrin *in* WY Cheung, ed, Calcium and Cell Function, Vol IV, Academic Press, New York, pp 152-174
12. **Nuccitelli R, Jaffe LF** (1976) Current pulses involving chloride and potassium efflux relieve excess pressure in *Pelvetia* embryos. Planta **131**: 315-320.
13. **Rudolph U, Gross H, Schnepf E** (1989) Investigations of the turnover of the putative cellulose-synthesizing particle "rosettes" within the plasma membrane of *Funaria hygrometrica* protonema cells. Protoplasma **148**: 57-80
14. **Speksnijder JE, Miller AL, Weisenseel MH, Chen T-H, Jaffe LF** (1989) Calcium buffer injections block fucoid egg development by facilitating calcium diffusion. Proc Natl Acad Sci USA **86**: 6607-6611
15. **Torrey JG, Galun E** (1970) Apolar embryos of *Fucus* resulting from osmotic and chemical treatment. Am J Bot **57**: 111-119

A MOLECULAR GENETICS AND MUTANT ANALYSIS APPROACH FOR ELUCIDATION OF MOLECULAR MECHANISMS OF CALCIUM SIGNAL TRANSDUCTION THROUGH CALMODULIN:CALMODULIN-BINDING PROTEIN COMPLEXES

Emily Wilson, D. Martin Watterson, and Mark Collinge

Department of Pharmacology, Vanderbilt University School of Medicine, Nashville, TN 37232, USA

External stimuli such as hormones, light, gravity, pressure, and temperature elicit immediate, as well as gradual, responses from eukaryotic cells. Many of the immediate responses are transient in nature, yet are key to the initiation or modulation of later events. One such response to extracellular stimuli, which appears to be common throughout nature, is a transient rise in intracellular-ionized calcium with maintenance, in most cases, of the total intracellular calcium concentration. The molecular mechanisms by which these increases in intracellular-ionized calcium are transduced into biological responses include interaction with calcium-binding proteins which are often components of larger supramolecular structures (reviewed in ref. 26). Calmodulin (CaM) is an ubiquitous eukaryotic protein that is an example of this class of calcium-binding protein. Early reports, based on activity analyses, postulated that there was a large molar excess production of CaM in eukaryotic cells (3). However, later studies showed CaM to be a unique chemical entity that is a component of a branched cellular regulatory system composed of a number of CaM-binding proteins including protein kinases, cytoskeletal elements, and membrane transport systems. The concentration of each of these CaM-regulated proteins is such that the total molar concentration of all CaM-binding proteins may approximate that of CaM (24).

The results, to date, raise the question of whether there is coupled regulation of the biosynthesis of CaM and its cellular target proteins. In other words, the molecular mechanism by which CaM-mediated pathways selectively transduce intracellular calcium signals into biological responses appears to be through the response characteristics of supramolecular structures that include CaM and a particular CaM-binding protein. Therefore, the physiological responses depend on the proper synthesis, assembly, and subcellular targeting of these signal-transduction complexes. However, our knowledge about how CaM and its binding proteins are encoded, synthesized, and assembled into functional complexes, capable of transducing calcium signals into biological responses, is incomplete and heterogeneous.

As schematically outlined in Figure 1, the major steps required for assembly of a CaM-regulated protein complex include transcription of the genes encoding both CaM and the CaM-binding protein, appropriate RNA processing, translation into protein products, post-translational modification, proper molecular recognition and assembly, and subcellular localization of the protein complex. One approach that is being used in attempts to enhance our understanding of this important process is mutant analysis. Past and current studies that employ some form of a mutant analysis approach can be arbitrarily grouped into four related categories: *i*) analysis and perturbation of the transcriptional units encoding CaM and CaM-binding proteins; *ii*) analysis of mutant organisms with defects in calcium-regulated responses that might be mediated by CaM; *iii*) *in vitro* site-directed mutagenesis of DNA-encoding CaM and CaM-binding proteins, combined with detailed *in vitro* functional analysis of purified proteins; and *iv*) *in vitro* mutagenesis of DNA-encoding CaM and CaM-binding proteins combined with *in vivo* phenotypic analysis of the mutant genes and protein. This report will briefly outline some studies in each of these categories and show how they contribute to our understanding of how CaM-mediated calcium signal transduction complexes are encoded, inherited, assembled, and function.

ANALYSIS AND PERTURBATION OF GENETIC UNITS

Disruption of the CaM gene in the yeasts *Saccharomyces cerevisiae* (5) and *Schizocharomyces pombe* (23) showed that the CaM gene product is essential for cell survival. However, the studies, to date, do not address whether this phenomenon is unique to yeast, whether just one branch of the CaM-regulatory system is "essential," or if it is the whole CaM-signaling system that is necessary. Similar studies in other organisms and gene-disruption experiments of specific CaM-binding proteins are two types of approaches that could provide additional insight. Alternatively, techniques such as the use of constructs that produce antisense RNA in response to an inducer might prove useful.

The gene-disruption studies in yeast, the fact that CaM is involved in several cellular regulatory pathways, and the former belief that all CaM-binding proteins use common structural features of CaM led to the generally held assumption that any mutation in CaM would be fatal. However, *in vitro* mutagenesis experiments clearly showed that mutations of CaM genes which altered the amino acid sequence of phylogenetically invariant regions had a selective regulatory effect (4, 10, 18). This demonstrated that it was possible to have genetic mutations with selective effects. Recently, a class of *Paramecium* behavioral mutants has been produced in which the inherited

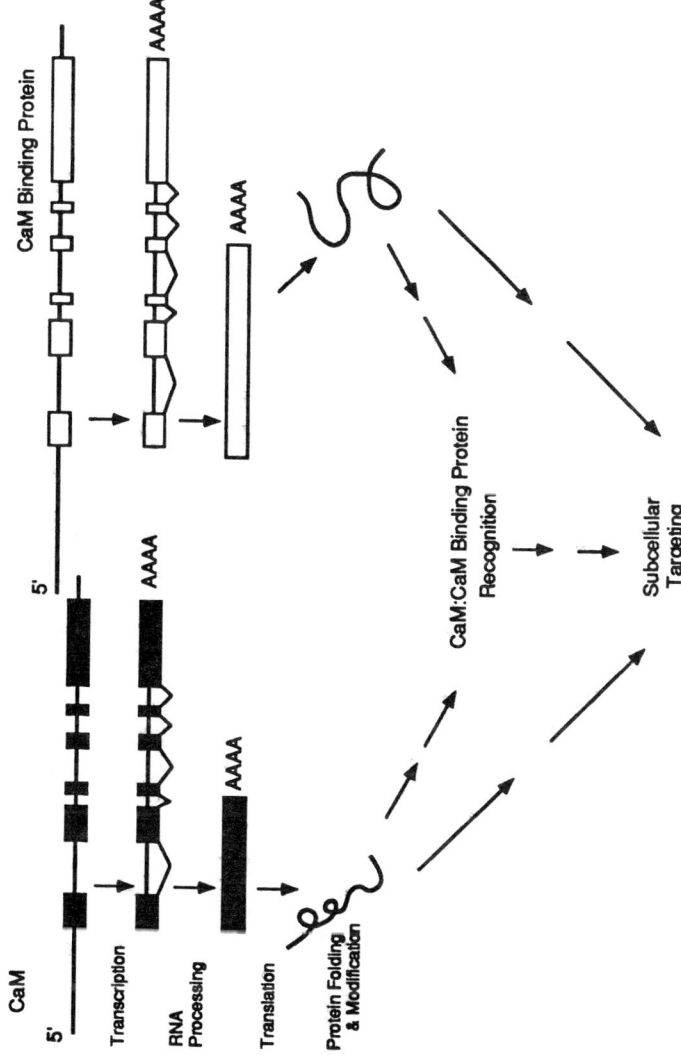

Figure 1. Diagrammatic representation of the steps involved in formation of CaM:CaM-binding protein complex. The steps include: RNA processing, protein modification, and folding of both CaM and CaM-binding proteins, followed by appropriate assembly and targeting of the complex.

defect has been localized to point mutations within CaM (13, 20). In this system, the CaM mutations are nonlethal and affect calcium modulated ion channel activity. These findings are a precedent for inherited mutations in CaM-gene exons being nonlethal as well as selective. Inherited mutations of protein complexes containing CaM-binding proteins have been described at the phenotypic level, but the molecular genetic bases of these nonlethal inherited mutations are still under investigation (21).

Clearly, it has now been demonstrated that it is possible to have nonlethal, inherited mutations of CaM:CaM-binding protein complexes which can lead to specific pathogenic states. Our current state of knowledge, however, is limited due to a lack of insight into how the eukaryotic cell is able to regulate the expression of the normal (wild-type) gene encoding CaM and CaM-binding proteins. Characterization of the transcriptional units for CaM and CaM-binding proteins is a necessary first step in attempts to elucidate molecular mechanisms involved in the expression of such genes. For CaM, the initial step has been achieved for several phylogenetically diverse species (5, 12, 14, 15, 23, 33). These studies have shown that, in contrast to the phylogenetic conservation of the amino acid sequence of CaM (19), there is extensive diversity in the number and genomic organization of CaM genes, with no clear phylogenetic trend.

In addition to a basal level of CaM production within all eukaryotic cells, there appears to be quantitative changes in mRNA and protein levels under some circumstances, including cell-cycle progression (1), transformation, and viral oncogene expression (2, 29). For example, Watterson *et al.* (29) demonstrated that viral oncogene expression altered the levels of a unique molecule that was later called CaM. These studies provided a precedent for CaM levels being subject to quantitative regulation. This concept was subsequently confirmed in another transformation system (2), and the mechanism for the increased CaM levels was shown (24, 32) to be due to a selective increase in CaM-specific mRNA levels. More recently, Zimmer and co-workers (33) have shown that deflagellation/stress stimulates a selective increase in CaM mRNA levels in the alga *Chlamydomonas reinhardtii*. Altogether, these results support a hypothesis that CaM levels are subject to quantitative regulated expression, as well as basal-level expression. However, the relative contributions of transcriptional regulation and mRNA stability to this regulated expression have not been elucidated.

Initial work on regulated expression of CaM genes raises the possibility that a portion of the regulation may be through transcriptional events, although direct evidence is lacking. For example, the 5'-untranslated region of several CaM genes (7, 14, 33) have sequence motifs that are indicative of regulated transcription and binding sites for well-characterized, DNA-binding proteins (31). One example is the *Chlamydomonas* CaM gene (33), which contains stress-response element motifs at appropriate distances from the transcription initiation site. In addition, Epstein *et al.* (7) have used the chimeric construct/reporter gene approach to test the ability of 5'-flanking regions of the chicken CaM gene to drive transcription of the chloramphenicol acetyltransferase (CAT) gene transfected into cultured cells.

Overall, the initial studies suggest that certain 5'-flanking regions may be important in regulation of CaM transcription. However, useful landmarks within the CaM-transcriptional unit vicinity, such as DNAse-hypersensitive sites, have not been described. Hypersensitive sites (6) are thought to represent regions of chromosomal DNA near sequences that bind transcriptional regulatory proteins and, therefore, these sequences may be involved in initiation or regulation of transcription. More detailed mapping of these regions can potentially identify sequences that bind these regulatory proteins. Other assays, such as gel shift assays, can be used to help in the identification of specific regulatory proteins. However, these studies are at an early phase, even for the analysis of basal-gene expression, and studies of regulated expression will be difficult with today's technology.

Our knowledge about gene organization and biosynthetic regulation of CaM-binding proteins is at a much earlier state than that for CaM. Unfortunately, at this stage, there is little evidence regarding the structural organization of genes encoding CaM-binding proteins. However, Van Eldik *et al.* (24) have shown that levels of some CaM-binding proteins are altered by viral oncogene expression. The protein and mRNA levels for one CaM-regulated protein, myosin light chain kinase (MLCK) (25, 27), were also shown to be altered by the same mechanisms that perturb CaM levels. These studies suggest that there is a coupling, at some level, of the biosynthesis and regulation of CaM and some of its binding proteins.

As additional information about genomic organization of genes encoding CaM-binding proteins and how their expression is altered by stimuli and cellular states becomes available, it should be possible to address the question of how a cell can regulate the synthesis and assembly of a CaM-regulated complex in a timely and correct manner. This knowledge is key to our understanding of how CaM-regulated protein complexes and supramolecular structure are made at the appropriate time and targeted to the appropriate place during cellular replication and differentiation in multicellular organisms.

ANALYSIS OF MUTANT ORGANISMS

Analysis of mutant organisms and cell lines has been useful for studying various signal transduction pathways. For the CaM system, it was previously thought that mutations in CaM would be lethal. Behavioral mutants of *Paramecium* have been described in which the defects have been localized to a single amino acid coding change in CaM (13, 20). These mutants have defects in regulation of calcium-dependent ion channel regulation; yet other CaM-modulated systems in the same organism are not altered. Two such CaM mutants have been described. These mutants have distinct amino acid coding changes and have similar, yet distinct, phenotypes. As previously mentioned, these studies demonstrated that mutations within CaM are possible and suggest that other such selective mutations of CaM may exist in nature.

Inherited mutations of CaM-dependent protein complexes and mutant organisms containing defects in CaM-binding proteins have been described. Several *Chlamydomonas* mutants have been described with defects in flagellar assembly. Initial analysis of some of these mutants suggest that there is alteration in some of the CaM-binding proteins (Van Eldik, unpublished data). Also, inherited diseases, such as McArdles Syndrome and glycogen storage disease, have been described at the phenotypic level as defects in phosphorylase kinase, a CaM-regulated system (21). Although the molecular mechanism for these defects has not been identified, it is clear that CaM-regulated protein complexes are involved in certain genetic disorders and that characterization of these mutations will aid in the understanding of calcium-mediated signal transduction.

IN VITRO MUTAGENESIS AND *IN VITRO* FUNCTIONAL ANALYSIS

Since Roberts *et al.* (18) introduced the use of recombinant DNA technology to the CaM system, a series of mutant CaMs have been produced that are isofunctional, have enhanced activity, or have diminished activity with one or more CaM-regulated enzymes (10). One notable trend with most of the mutations that have decreased activity with a given CaM-regulated enzyme is the selectivity of the effect, *i.e.* all CaM-regulated enzymes are not affected by a given CaM mutation. For example, CaMs have been engineered that differentially affect MLCKs (4, 28), the transport calcium ATPase (11), and NAD kinase (4). Another utility of *in vitro* mutagenesis of CaM has been to elucidate specific features of CaM that are important in protein:protein interactions, or protein recognition. For example, Weber *et al.* (30) showed that certain mutations of CaM affect the negative surface charge properties of CaM and selectively alter the interaction of CaM with two CaM-regulated protein kinases. These negative charges, or electrostatic forces, may play an important role in recognition and assembly of specific CaM-regulated protein complexes.

Analogous mutagenesis studies are emerging with some of the CaM-dependent protein kinases. Specifically, several deletion mutants of MLCKs have been reported (28) and a site-specific mutant of CaM kinase II has been reported (9). The deletion studies have generated kinases that are CaM-independent and thus are constitutively active. It is foreseen that mutant kinases could be constructed in which the CaM-binding region has been altered, the autoinhibitory region altered, the catalytic region altered, or substrate binding site altered. These studies will not only provide knowledge about the molecular mechanism of CaM-dependent protein kinases, but will also complement the studies done with mutant CaMs. This type of mutant analysis approach has the strength of allowing a large number of mutants to be analyzed rapidly and quantitatively, the potential for designing and producing mutations with desired effects, and the potential for analyzing, at a molecular level, specific portions of the proteins that are involved in protein:protein interactions, subcellular targeting, and intramolecular interactions.

IN VIVO PERTURBATION OF THE CaM SYSTEM

This general type of study incorporates the strengths of the site-specific mutagenesis approach with that of mutant organism analysis. Mutant cells or organisms can be made in which the concentration of CaM or specific CaM-binding proteins (*e.g.* protein kinase activities) are quantitatively altered either by overexpression of specific protein products or by decreasing the expression with antisense RNA production. Although overexpression is a useful approach, any phenotypic effects observed must be characterized to determine if they are a reflection of perturbed regulation, toxicity, or competition with other calcium regulatory systems. Another type of study that might help to alleviate these concerns would be expression of mutant proteins that have been well characterized *in vitro* and shown to perturb specific branches of the CaM-regulatory system.

A variety of experimental systems could be used for this type of study with each having its particular strengths and weaknesses. Unicellular organisms, such as yeast, *Chlamydomonas*, and *Paramecium*, are exceptionally appealing for such studies because they have been well defined both biochemically and genetically. Both yeast (5, 23) and *Chlamydomonas* (33) CaM are encoded by a single gene, so gene disruption experiments coupled with expression of a transfected mutated gene may be possible. Recently, the feasibility for affecting specific pathways has already been demonstrated. Takeda and co-workers (22) showed in yeast that expression of a transfected CaM with a single amino acid coding change has specific physiological effects. While these early phylogenetic systems are excellent for initial studies, these species appear to lack some of the better-characterized, CaM-regulatory systems found in higher organisms and may not exhibit the complex regulatory systems found in higher plants and animals. Therefore, there is a need for similar studies in multicellular organisms representative of later phylogenetic species.

Cultured vertebrate cell lines provide some of the technical appeal of unicellular systems, but maintain much of signal transduction mechanisms and cellular responses to external stimuli found in vertebrates. This offers the opportunity of perturbing cellular physiology by altering CaM and CaM-binding protein activity under a comparatively controlled environment. Rasmussen *et al.* (16) showed that overexpression of CaM has effects on cell cycle regulation. This type of study, combined with knowledge gained from virus-transformed cells, suggests that quantitative regulation of CaM and its target proteins is important for normal cellular growth and development. Means and co-workers (8) have also recently shown the feasibility of utilizing the technology of transgenic animals to specifically overexpress CaM in insulin-producing islet cells. They suggest that inappropriate regulation may lead to disease states. As noted earlier, it is difficult to dissociate phenotypic responses seen with overexpression from toxicity/stress and from competition with other calcium-modulated pathways not utilizing CaM. Similar studies utilizing CaM mutants that have been shown *in vitro* to affect specific enzymes or overexpression of CaM-regulated enzymes may help dissect the role of specific pathways in normal cellular function. For example, overexpression of CaMs that specifically affect MLCK activation can be complemented with the regulated expression of MLCK constructs that are CaM-independent.

Higher plants are appealing for these transfection studies. As with animals, there is the potential for perturbing cellular physiology in cell culture and by making transgenic plants. Undifferentiated plant cells have the potential for developing into a variety of cell types making plants a good system for studying the role of CaM and CaM-binding proteins in cellular differentiation. CaMs that preferentially alter plant NAD kinase activity have been produced by site-specific mutagenesis, and vectors capable of allowing selected gene expression in higher plants are available (4, 17). These mutant CaM will be interesting to express in a plant system in order to better understand the role of this enzyme *in vivo*. Overall, the use of well-characterized CaM mutants in plant systems appears to provide the best opportunity for elucidating how CaM may be involved in development. In summary, the knowledge and mechanistic insight obtained from *in vitro* site-specific mutagenesis studies, combined with the availability of transfected eukaryotic cells from a broad spectrum of phylogenetic species, may aid in the development of rational approaches for the study of cellular perturbation by components of CaM regulatory systems.

ACKNOWLEDGMENTS

This research was supported in part by National Institutes of Health Grant GM30861 (DMW). We thank Janis Elsner for her help with the preparation of this manuscript.

LITERATURE CITED

1. Chafouleas JG, Bolton WE, Hidaka H, Boyd AE, Means AR (1982) Calmodulin and the cell cycle: involvement in regulation of cell-cycle progression. Cell 28: 41-50
2. Chafouleas JG, Pardue RL, Brinkley BR, Dedman JR, Means AR (1981) Regulation of intracellular levels of calmodulin and tubulin in normal and transformed cells. Proc Natl Acad Sci USA 78: 996-1000
3. Cheung WY (1980) Calmodulin plays a pivotal role in cellular regulation. Science 207: 19-27
4. Craig TA, Watterson DM, Prendergast FG, Haiech J , Roberts DM (1987) Site-specific mutagenesis of the α-helices of calmodulin. Effects of altering a charge cluster in the helix that links the two halves of calmodulin. J Biol Chem 262: 3278-3284
5. Davis TN, Urdea MS, Masiarz FR, Thorner J (1986) Isolation of the yeast calmodulin gene: calmodulin is an essential protein. Cell 47: 423-431

6. **Elgin SCR** (1988) The formation and function of DNase I hypersensitive sites in the process of gene activation. J Biol Chem **263**: 19259-19262
7. **Epstein PN, Christenson MA, Means AR** (1989) Chicken calmodulin promoter activity in proliferating and differentiated cells. Mol Endocrinol **3**: 193-202
8. **Epstein PN, Overbeek PA, Means AR** (1989) Calmodulin-induced early-onset diabetes in transgenic mice. Cell **58**: 1067-1073
9. **Fong Y-L, Taylor WL, Means AR, Soderling TR** (1989) Studies of the regulatory mechanism of Ca^{2+}/calmodulin-dependent protein kinase II. Mutation of threonine 286 to alanine and aspartate. J Biol Chem **264**: 16759-16763
10. **Haiech J, Watterson DM** (1988) Site-specific mutagenesis and protein engineering approach to the molecular mechanism of calcium signal transduction by calmodulin. *In* Ch Gerday, R Gilles, L Bolis, eds, Calcium and Calcium Binding Proteins. Springer-Verlag, Berlin, Heidelberg, pp 191-200
11. **Krebs J, Vorherr T, James P, Carafoli E, Craig TA, Watterson DM** (1989) Structural details of the interaction of calmodulin with the plasma membrane Ca^{2+}-ATPase. *In* R Pochet, DEM Lawson, CW Heizmann, eds, Proceedings of First European Symposium on Calcium Binding Proteins. Plenum Press, New York, pp 163-167
12. **LeJohn HB** 1989 Structure and expression of fungal calmodulin gene. J Biol Chem **264**: 19366-19372
13. **Lukas TJ, Wallen-Friedman M, Kung C, Watterson DM** (1989) *In vivo* mutations of calmodulin: A mutant *Paramecium* with altered ion current regulation has an isoleucine-to-threonine change at residue 136 and an altered methylation state at lysine residue 115. Proc Natl Acad Sci USA **86**: 7331-7335
14. **Nojima H** (1989) Structural organization of multiple rat calmodulin genes. J Mol Biol **208**: 269-282
15. **Nojima H, Sokabe H** (1987) Structure of a gene for rat calmodulin. J Mol Biol **193**: 439-445
16. **Rasmussen CD, Means AR** (1989) Calmodulin is required for cell-cycle progression during G1 and mitosis. EMBO J **8**: 73-82
17. **Roberts DM, Burgess WH, Watterson DM** (1984) Comparison of the NAD kinase and myosin light chain kinase activator properties of vertebrate, higher plant, and algal calmodulins. Plant Physiol **75**: 796-798
18. **Roberts DM, Crea R, Malecha M, Alvarado-Urbina G, Chiarello RH, Watterson DM** (1985) Chemical synthesis and expression of a calmodulin gene designed for site-specific mutagenesis. Biochemistry **24**: 5090-5098
19. **Roberts DM, Lukas TJ, Watterson DM** (1986) Structure, function, and mechanism of action of calmodulin. CRC Critical Rev in Plant Sci **4**: 311-339
20. **Schaefer WH, Hinrichsen RD, Burgess-Cassler A, Kung C, Blair IA, Watterson DM** (1987) A mutant *Paramecium* with a defective calcium-dependent potassium conductance has an altered calmodulin: A nonlethal selective alteration in calmodulin regulation. Proc Natl Acad Sci USA **84**: 3931-3935
21. **Scriver CR, Beaudet AL, Sly WS, Valle D** (1989) The metabolic basis of inherited disease. McGraw-Hill, NY, pp 1-1476
22. **Takeda T, Imai Y, Yamamoto M** (1989) Substitution at position 116 of *Schizosaccharomyces pombe* calmodulin decreases its stability under nitrogen starvation and results in a sporulation-deficient phenotype. Proc Natl Acad Sci USA **86**: 9737-9741
23. **Takeda T, Yamamoto M** (1987) Analysis and *in vivo* disruption of the gene coding for calmodulin in *Schizosaccharomyces pombe*. Proc Natl Acad Sci USA **84**: 3580-3584

24. Van Eldik LJ, Burgess WH (1983) Analytical subcellular distribution of calmodulin and calmodulin-binding proteins in normal and virus-transformed fibroblasts. J Biol Chem 258: 4539-4547
25. Van Eldik LJ, Watterson DM, Burgess WH (1984) Immunoreactive levels of myosin light-chain kinase in normal and virus-transformed chicken embryo fibroblasts. Mol Cell Biol 4: 2224-2226
26. Van Eldik LJ, Zendegui JG, Marshak DR, Watterson DM (1982) Calcium-binding proteins and the molecular basis of calcium action. Int Rev Cytol 77: 1-61
27. Van Eldik LJ, Zimmer WE, Barger SW, Watterson DM (1989) Perturbation of the calmodulin system in transformed cells. In R Pochet, DEM Lawson, CW Heizmann, eds, Calcium Binding Proteins in Normal and Transformed Cells. Plenum Press, NY, pp 111-120
28. Watterson DM, Haiech J, Zimmer WE, Shattuck R, Shoemaker M, Lau W, Craig T, Lukas T (1988) The molecular biology of a calmodulin regulated protein kinase system. CSH Symposium on Quantitative Biology. 53: 185-193
29. Watterson DM, Van Eldik LJ, Smith RE, Vanaman TC (1976) Calcium-dependent regulatory protein of cyclic nucleotide metabolism in normal and transformed chicken embryo fibroblasts. Proc Natl Acad Sci USA 73: 2711-2715
30. Weber PC, Lukas TJ, Craig TA, Wilson E, King MM, Kwiatkowski AP, Watterson DM (1989) Computational and site-specific mutagenesis analyses of the asymmetric charge distribution on calmodulin. Proteins: Structure, Function & Genetics 6: 70-85
31. Yamamoto KR (1989) A conceptual view of transcriptional regulation. Am Zool 29: 537-547
32. Zendegui JG, Zielinski RE, Watterson DM, Van Eldik LJ (1984) Biosynthesis of calmodulin in normal and virus-transformed chicken embryo fibroblasts. Mol Cell Biol 4 : 883-889
33. Zimmer WE, Schloss JA, Silflow CD, Youngblom J, Watterson DM (1988) Structural organization, DNA sequence and expression of the calmodulin gene. J Biol Chem 263: 19370-19383

LOCALIZATION OF CALCIUM-BINDING PROTEINS IN DIVIDING PLANT CELLS

Susan M. Wick

Department of Plant Biology, University of Minnesota, St. Paul, MN 55108, USA

Considering the increasing numbers and types of calcium-binding proteins that have been characterized in recent years (16), there probably are few regions of the plant cell that do not contain one or more of them. Calcium-binding proteins involved in regulation of reactions that occur throughout the soluble cytoplasm are likely to have widespread and somewhat uniform distribution. In addition to these, it is of interest to know whether there are other calcium-binding proteins whose spatial distribution reflects the polarity of cells within an organized tissue, or reflects localized activation of a Ca^{2+}-dependent process. One aspect of the regulation of plant cellular morphogenesis is control of the orientation of cell division, which is influenced in some cells by the location and orientation of mitosis (6). Both mitosis and cytokinesis involve extensive rearrangement of cytoskeletal elements, which are known to be sensitive to levels of Ca^{2+} (reviewed in 18). In the plant mitotic apparatus, ER elements in the form of fenestrated lamellae and a tubular-reticulate network, both shown by various methods to sequester Ca^{2+}, cap the spindle poles and interdigitate among kinetochore microtubules, respectively (9). The question of whether calcium-binding proteins have a particular localization relative to the cytoskeleton and membrane system of the mitotic apparatus and cytokinetic apparatus (phragmoplast) is beginning to be addressed (18).

Two general categories of calcium-binding proteins that can be examined with respect to their location during mitosis and cytokinesis are calcium-storage proteins and regulatory calcium-binding proteins. In the former category is calsequestrin, the principal calcium-storage protein in the lumen of sarcoplasmic reticulum, the specialized, smooth ER of muscle cells. Recently, a protein resembling calsequestrin has been found in the ER lumen of sea urchin cells (8), and the possibility that calsequestrin may be commonly distributed in other nonmuscle cells is likely. Other Ca^{2+}-storage proteins that merit investigation in plant cells are the reticuloplasmins. Defined as proteins within the lumen of the ER, they are found, to date, in a variety of animal cells (12). At least three of the reticuloplasmins are able to bind and store Ca^{2+}: endoplasmin, protein disulphide isomerase, and a 55-kD protein (12).

Regulatory calcium-binding proteins include the calcimedins (13) and the extensively studied protein family that includes calmodulin. A role often suggested for calmodulin is that of mediator of Ca^{2+}-dependent microtubule depolymerization, *i.e.* as a regulator of spindle function (reviewed in 5, 15, 17-19). Two other members of the calmodulin family are the yeast *CDC31* gene product (2, 11), which is required for spindle pole body duplication, and the contractile protein centrin (caltractin) (11), which is associated with basal rootlets of algal cells (14) and centrosomes of algal (14) and mammalian (1) cells. In this paper, I will briefly summarize and compare results concerning immunofluorescence localization of two calcium-binding proteins, calmodulin and a centrin-like protein, in dividing plant cells.

MATERIALS AND METHODS

Cells from root tips of *Pisum sativum* and *Allium cepa* L. seedlings were prepared for immunofluorescence microscopy according to procedures detailed in Wick *et al.* (19) for optimal preservation of calcium-binding antigens, *i.e.* eliminating EGTA from any of the processing steps. Calmodulin localization was obtained with rabbit anti-spinach calmodulin and either fluorescein- or rhodamine-conjugated secondary antibodies. For detection of centrin-like antigens, Dr. Jeffrey Salisbury, Mayo Clinic Foundation, generously supplied rabbit antibodies to centrin from the alga *Tetraselmis striata*, which were used with secondary antibodies as above. DNA was labeled with the fluorescent indicator Hoechst 33258.

RESULTS AND DISCUSSION

Calmodulin Localization

Details of calmodulin localization in meristematic cells have been reported previously (18, 19). From the beginning of mitosis through cytokinesis, the bulk of the cellular calmodulin is found co-localized with microtubules (Figs. 1, 2, and 5). The distribution of calmodulin in mitotic root meristem cells, as visualized with immunofluorescence microscopy, looks identical to that found in endosperm cells (17), in which immunoelectron microscopy indicates specific labeling of kinetochore bundles of microtubules (17). Information on the dynamics of a fluorescent calmodulin analogue microinjected into mitotic mammalian cells provides further evidence that calmodulin in the spindle is bound to microtubules, and that its accumulation in the spindle depends on the presence of microtubules (15). The normal progression of mitosis is impeded when calmodulin antagonists, calmidazolium (17) or 48/80 (19), are applied, suggesting that calmodulin in the spindle does exert a regulatory role. Recent evidence indicates that the mode of action of

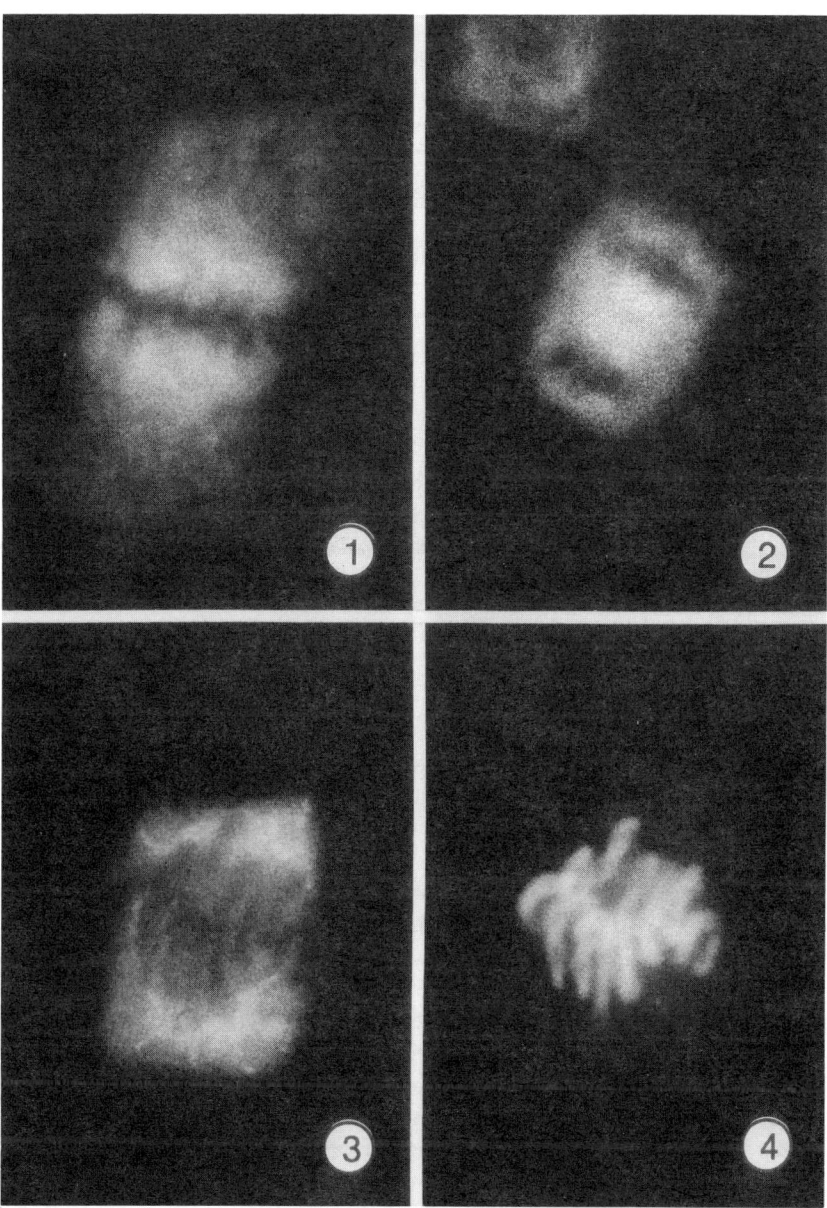

Figures 1-4. Figure 1. Localization of calmodulin in the mitotic spindle of a *P. sativum* root tip cell. Dark region in center of cell is area occupied by metaphase chromosomes. X 1800. **Figure 2.** Localization of calmodulin in phragmoplast during cytokinesis, *P. sativum* root tip cell. Dark regions of cell are the two sets of recently-separated daughter chromosomes. X 1300. **Figure 3.** Localization of centrin-like protein around spindle pole region in mitotic *A. cepa* root tip cell. X 1200. **Figure. 4.** Corresponding DNA image, same cell as in Figure 3. X 1200.

Ca^{2+}-calmodulin in the spindle is to activate a protein kinase that phosphorylates a 62-kD protein, also localized on the microtubules, thereby resulting in microtubule depolymerization (5). Concentrating a calcium-binding regulatory protein specifically along the microtubules that need to shorten during anaphase may provide the cell with a method that is more spatially accurate than would be possible by trying to regulate microtubule depolymerization solely on the basis of highly localized releases of Ca^{2+}.

Bundles or dense arrays of microtubules found at other times of the cell cycle, e.g. the preprophase band, do not have corresponding accumulations of calmodulin (19), nor does injected calmodulin bind to taxol-bundled interphase microtubules (15), suggesting that not all microtubules have the capacity to associate with calmodulin. In terms of morphogenesis, this also suggests that calmodulin is not involved in establishing the site and polarity of cell division, which are determined by the time the preprophase band appears (6). (However, calmodulin distribution does reflect the intracellular polarity that is manifested when spindle poles are defined early in mitosis.) For reasons not obvious at this time, the phragmoplast of endosperm cells does not show an accumulation of calmodulin (17), whereas the phragmoplast of meristematic cells does (18, 19).

Other systems in which localized concentrations of calmodulin are found to reflect cellular polarity include tip-growing fungal, flowering plant, moss (7), and algal (3, 7) cells, and root cap columella cells (4).

Localization of a Centrin-Like Protein

Algal centrin, known also as caltractin (11), has been extensively characterized (reviewed in 14) and is proposed to function both as part of a microtubule organizing center (MTOC) (1, 14) and as a contractile organelle (10, 14). Antibodies to algal centrin immunoprecipitate a Ca^{2+}-binding protein from *A. cepa* root tips (unpublished results). When used for immunofluorescence microscopy, these same antibodies also reveal the location of the centrin-like antigen in root meristem cells (Figs. 3, 4, and 5). The antigen is localized in the general region of the nuclear envelope and perinuclear ER throughout most of the cell cycle, becoming redistributed at mitosis to spindle pole regions, where it appears to have a distribution similar to that of the ER that surrounds the poles and invades the spindle region along kinetochore microtubule bundles (9) (manuscript in preparation). At cytokinesis, it associates with the reforming nuclear envelope of each of the daughter cells.

In its postulated role as a MTOC, algal centrin is implicated in organization of the flagellar basal apparatus, while its contractile function is manifested in such activities as movements of the nucleus and spindle poles, positioning of centrioles at mitosis, and reorientation of basal bodies (14). It is also involved in contraction of striated fibers associated with the contractile transverse flagellum of some dinoflagellates (10). Throughout the cell cycle of *Chlamydomonas*, in spite of considerable changes in its distribution, centrin always remains associated with the nuclear envelope (14). Identification of a

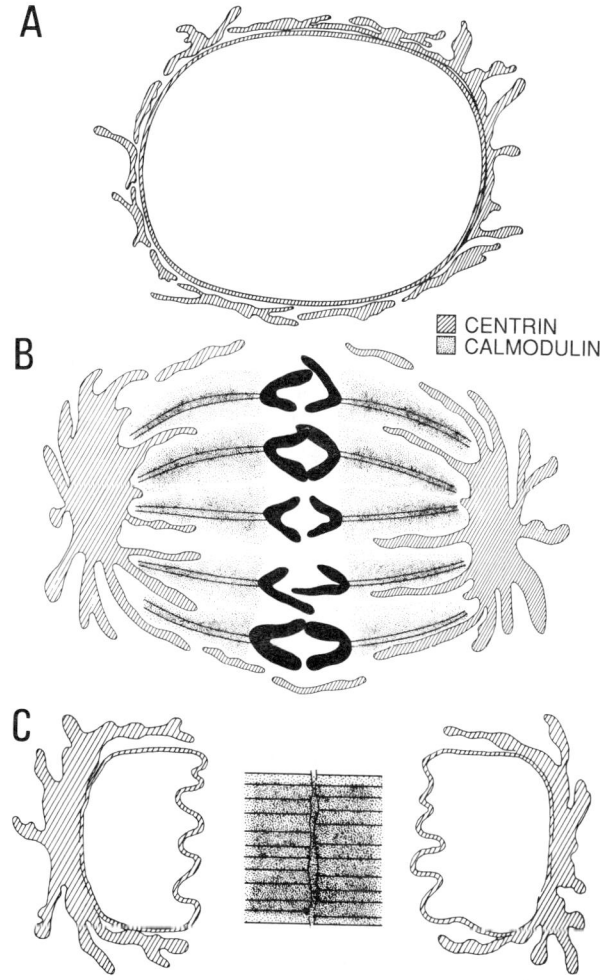

Figure 5. Distribution of calmodulin and centrin-like protein in meristematic cells. A = interphase, B = mitosis, C = cytokinesis.

centrosome-associated homologue of centrin that forms a pericentriolar lattice near the nucleus in mammalian cells (1) raises the question of whether proteins related to centrin are ubiquitous. The intriguing possibility that the centrin-like protein in plant cells is the functional counterpart of the nuclear envelope- and centrosome-associated contractile centrin of algal cells remains to be tested.

CONCLUSIONS

Figure 5 summarizes the distribution of the two calcium-binding proteins, calmodulin and the centrin-like protein, through interphase, mitosis, and cytokinesis in meristematic plant cells. The two proteins display distinctly different localization patterns, presumably indicating that each is involved in different regulatory and/or structural activities. While the function of calmodulin in animal cell mitosis and the functions of centrin in algal cells throughout the cell cycle are becoming more clear, a great number of questions remain concerning what these proteins are doing in plant cells. If the centrin-like protein has a function analogous to the ones it has in algae, perhaps it is involved in contraction of the ER to the poles at the time spindle polarity is established.

Whatever the functions of these two proteins, their different distributions may be an indication of one way in which a cell deals with the need to spatially and temporally regulate a large number of Ca^{2+}-modulated activities: rather than relying entirely on very highly localized differences in Ca^{2+} to simultaneously regulate several events, it appears that the cell can distribute differentially some of its regulatory calcium-binding proteins, each of which could, in the proper environment, regulate a particular cell function.

ACKNOWLEDGMENTS

The author gratefully acknowledges support through USDA 87-CRCR-1-2530 and the Graduate School of the University of Minnesota, and the gift of anti-centrin from J. Salisbury. The insightful comments of Franklin Harold were much appreciated, as well.

LITERATURE CITED

1. **Baron AT, Salisbury JL** (1988) Identification and localization of a novel, cytoskeletal, centrosome-associated protein in PtK$_2$ cells. J Cell Biol **107**: 2669-2678
2. **Baum P, Furlong C, Byers B** (1986) Yeast gene required for spindle pole body duplication: Homology of its product with Ca^{2+}-binding proteins. Proc Natl Acad Sci USA **83**: 5512-5516
3. **Cotton G, Vanden Driessche T** (1987) Identification of calmodulin in *Acetabularia*: its distribution and physiological significance. J Cell Sci **87**: 337-347
4. **Dauwalder M, Roux SJ, Hardison L** (1986) Distribution of calmodulin in pea seedlings: Immunocytochemical localization in plumules and root apices. Planta **168**: 461-470
5. **Dinsmore JH, Sloboda RD** (1989) Microinjection of antibodies to a 62 kD mitotic apparatus protein arrests mitosis in dividing sea urchin embryos. Cell **57**: 127-134
6. **Gunning BES, Wick SM** (1985) Preprophase bands, phragmoplasts, and spatial control of cytokinesis. J Cell Sci Suppl **2**: 157-179
7. **Haußer I, Herth W, Reiss H-D** (1984) Calmodulin in tip-growing plant cells, visualized by fluorescing calmodulin-binding phenothiazines. Planta **162**: 33-39

8. **Hensen JH, Begg DA, Beaulieu SM, Fishkind DJ, Bonder EM, Terasaki M, Lebeche D, Kaminer B** (1989) A calsequestrin-like protein in the endoplasmic reticulum of the sea urchin: Localization and dynamics in the egg and first cell cycle embryo. J Cell Biol 109: 149-161
9. **Hepler PK, Wick SM, Wolniak SM** (1981) The structure and role of membranes in the mitotic apparatus. *In* HG Schweiger, ed, International Cell Biology 1980-1981. Springer-Verlag, Berlin, pp 673-686
10. **Höhfeld I, Otten J, Melkonian M** (1988) Contractile eukaryotic flagella: Centrin is involved. Protoplasma 147: 16-24
11. **Huang B, Mengersen A, Lee VD** (1988) Molecular cloning of cDNA for caltractin, a basal body-associated Ca^{2+}-binding protein: Homology in its protein sequence with calmodulin and the yeast *CDC31* gene product. J Cell Biol 107: 133-140
12. **Macer DRJ, Koch GLE** (1988) Identification of a set of calcium-binding proteins in reticuloplasm, the luminal content of the endoplasmic reticulum. J Cell Sci 91: 61-70
13. **Moore PB, Kraus-Friedmann N, Dedman JR** (1984) Unique calcium-dependent hydrophobic binding proteins: possible independent mediators of intracellular calcium distinct from calmodulin. J Cell Sci 72: 121-133
14. **Salisbury JL** (1989) Algal centrin: Calcium-sensitive contractile organelles. *In* AW Coleman, LJ Goff, JR Stein-Taylor, eds, Algae as Experimental Systems. Alan R. Liss, NY, pp 19-37
15. **Stemple DL, Sweet SC, Welsh MJ, McIntosh JR** (1988) Dynamics of a fluorescent calmodulin analog in the mammalian mitotic spindle at metaphase. Cell Motil Cytoskeleton 9: 231-242
16. **Thompson MP, ed** (1988) Calcium-binding Proteins, Vol I and II. CRC Press, Boca Raton, FL
17. **Vantard M, Lambert A-M, De Mey J, Picquot P, Van Eldik LJ** (1985) Characterization and immunocytochemical distribution of calmodulin in higher plant endosperm cells: Localization in the mitotic apparatus. J Cell Biol 101: 488-499
18. **Wick SM** (1988) Immunolocalization of tubulin and calmodulin in meristematic plant cells. *In* MP Thompson, ed, Calcium-binding Proteins, Vol II. CRC Press, Boca Raton, FL, pp 21-45
19. **Wick SM, Muto S, Duniec J** (1985) Double immunofluorescence labeling of calmodulin and tubulin in dividing plant cells. Protoplasma 126: 198-206

VOLTAGE-DEPENDENT ACTIVATION OF CA^{2+}-REGULATED ANION CHANNELS AND K^+ UPTAKE CHANNELS IN *VICIA FABA* GUARD CELLS

Julian I. Schroeder and Susumu Hagiwara*

*Department of Physiology, UCLA School of Medicine,
Los Angeles, CA 90024, USA*

Calcium controls a variety of ion transport-associated processes in higher plant cells (5). For example, during stomatal movements, Ca^{2+} triggers closing of stomata and inhibits stomatal opening (2, 8, 15). Using patch-clamp techniques, we have recently found that elevation of cytosolic Ca^{2+} to the micromolar concentration leads to activation of a voltage-dependent anion conductance (12) and inhibition of K^+ uptake channels (12). This Ca^{2+}-induced modulation of voltage-dependent anion and K^+ channels provides molecular mechanisms for the Ca^{2+}-dependent regulation of stomatal movements by physiological signals. The voltage-dependent activation of Ca^{2+}-regulated anion channels and K^+ uptake channels in guard cells from *Vicia faba* will be investigated in this report.

MATERIALS AND METHODS

Guard cell protoplasts were isolated by modification of a standard procedure (3, 11). Patch clamp techniques (4) were applied to guard cell protoplasts as described in previous reports (11, 14). Data acquisition and analysis were performed by digitization and processing on a PDP 11/73 computer (11, 14).

RESULTS

In previous studies, we have shown that two major types of voltage-dependent K^+ channels prevail in the plasma membrane of guard cells (Fig. 1A; 11, 13, 14). One K^+ channel type is activated by depolarization to potentials more positive than the K^+ equilibrium potential and allows K^+ efflux, required for stomatal closing (1, 14). The other K^+ channel type is

---

*Deceased

activated at membrane potentials more negative than -90 mV and provides the molecular pathway for K^+ uptake during stomatal opening (11, 13, 14).

The question whether cytoplasmic Ca^{2+} regulates these ion transport mechanisms can be directly addressed by patch clamp techniques. The whole-cell patch clamp technique allows control of the cytoplasmic-free Ca^{2+} concentration ($[Ca^{2+}]cyt$), while monitoring of the effect of changes in cytoplasmic Ca^{2+} on ion channels in the guard cell plasma membrane (12). When $[Ca^{2+}]cyt$ was buffered to approximate resting values in the range of 0.1 µM, typical voltage-dependent inward and outward conducting K^+ channel currents were recorded (Fig. 1A; 11, 14). The inward conducting K^+ channels were activated at potentials negative to -90 mV at resting $[Ca^{2+}]cyt$ (Fig. 1A; 11).

Raising $[Ca^{2+}]cyt$ to elevated values of 1.5 µM resulted in several dramatic changes in ion channels of the guard cell membrane (Fig. 1B). The inward conducting K^+ channels were inhibited in the normal activation range. This inhibition of K^+ uptake channels is mediated by shifting the activation

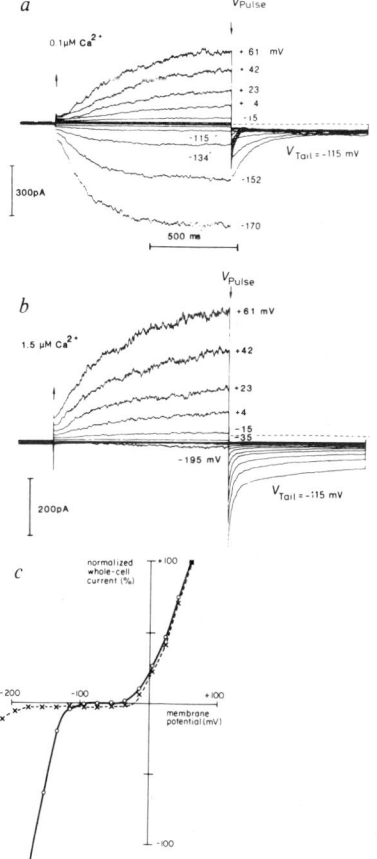

Figure 1. Effects of cytosolic Ca^{2+} on guard cell ion channels recorded with the whole-cell patch clamp technique. A, Voltage-dependent activation of inwardly conducting K^+ channel currents (downward deflections) and outwardly conducting K^+ channel currents (upward deflections) measured with 0.1 µM Ca^{2+} in the cytoplasm. B, Voltage-dependent anion and K^+ release currents measured with 1.5 µM Ca^{2+} in the cytoplasm. C, Current-voltage relationships of cells shown in (A) and (B). (Reprinted from 12 with permission of **Nature** Vol 338: 427-430 1989, Macmillan Magazines LTD.)

potential of these K^+ channels to strongly negative potentials of approximately -195 mV (Fig. 1C, 12).

Elevation of the cytoplasmic-free Ca^{2+} concentration to 1.5 μM produced further effects on guard cell ion channels. Voltage steps to potentials more positive than -75 mV elicited rapidly activating currents, in addition to the slowly activating K^+ channels at low $[Ca^{2+}]cyt$ (Fig. 1B). Furthermore, repolarization of the membrane to -115 mV (V_{Tail}) evoked slow relaxation currents (Fig. 1B). In recordings of the guard cell membrane potential, we found that these Ca^{2+}-activated voltage-dependent currents can depolarize guard cells to positive potentials (12). This finding shows that the voltage-dependent Ca^{2+}-activated currents are not solely carried by K^+-selective channels. Detailed studies, in which the ionic gradients across the plasma membrane were varied, show that these Ca^{2+}-activated ion channels have a permeability to anions (12).

The steep physiological gradient of free anions across the plasma membrane of guard cells and other higher plant cells would induce depolarization when anion permeable channels open. The Ca^{2+}-activation of voltage-dependent anion channels in guard cells provides a molecular mechanism by which Ca^{2+}-dependent depolarization of higher plant cells may be achieved.

Elevation of cytosolic Ca^{2+} activates anion permeable channels when the membrane is depolarized to potentials more positive than approximately -80 mV (Fig. 1B; 6, 12). In addition, Ca^{2+} activates slow relaxation currents when the membrane is stepped from depolarized potentials to more negative potentials (12; *e.g.* Fig. 1B, V_{Tail} = -115 mV). These slow relaxation currents may reflect hyperpolarization-induced closing ("deactivation") of Ca^{2+}-activated channels. This suggestion was examined by voltage pulse procedures.

Figure 2A (top) shows that hyperpolarization from a holding potential (V_H) of -35 to -155 mV leads to a large inward current which relaxes slowly. Subsequently, the membrane was held at a potential of V_H = -95 mV, a potential at which most anion channels are closed (Fig. 2A, middle). Hyperpolarization from -95 to -155 mV induces a much smaller inward current and no significant slow relaxation current. Subsequently, the membrane was again held at -35 mV and polarized to -155 mV (Fig. 2A, bottom trace), showing that the large inward current and the large relaxation current reappear when the holding potential is depolarized.

These data, along with other pulse protocols (Schroeder and Hagiwara, unpublished), indicate that the slow relaxation currents at negative potentials can be attributed to closing (deactivation) of Ca^{2+}-activated channels by hyperpolarization. From these results and analysis of steady-state anion-permeable currents, the steady-state voltage dependence of Ca^{2+}-activated anion currents can be deduced (Fig. 2B). Figure 2B shows the steady-state voltage dependence of Ca^{2+}-activated anion currents which were recorded by suppressing K^+ channel currents by use of Cs^+ ions (11). The steady-state voltage dependence of the Ca^{2+}-activated, anion-permeable conductance (Fig.

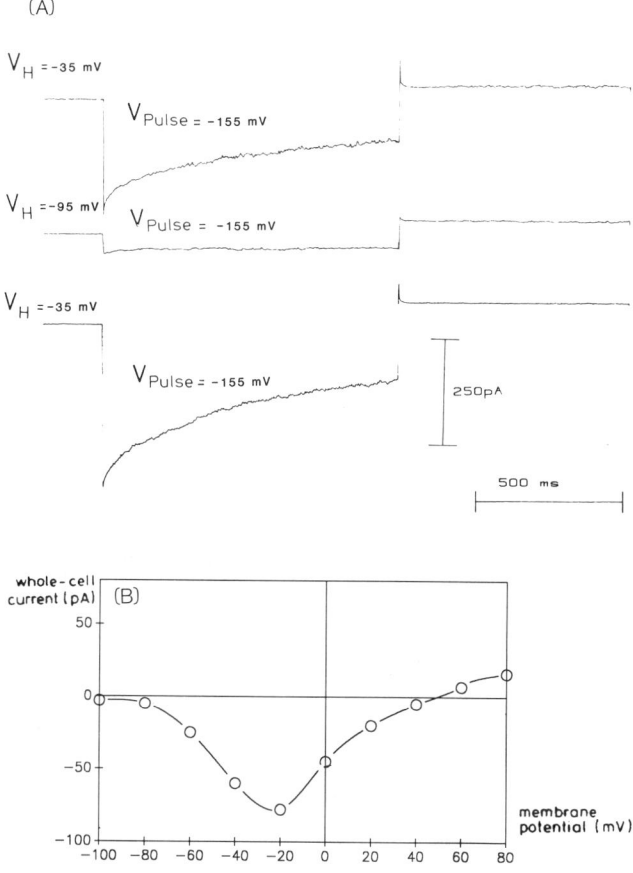

Figure 2. Voltage dependence of Ca^{2+}-regulated ion currents. A, The magnitude of the hyperpolarization-induced slow relaxation current depends strongly on how depolarized the holding potential is (V_H = -35 mV in top and bottom trace and V_H = -95 mV in the middle trace). These data indicate that slow relaxation currents correspond to closing (deactivation) of depolarization activated currents (see text). B, Steady-state current-voltage curve of Ca^{2+}-activated anion currents.

2B) is similar to the voltage dependence of anion-selective single channels in guard cells as characterized in detail by Keller *et al.* (6).

The Ca^{2+}-activated anion channels diminish within 2 to 10 min of whole-cell patch clamp recordings. Preliminary results suggest that this time-dependent reduction of anion channels can be attributed to washout of essential components in the cytoplasm during whole-cell dialysis. Washout of Ca^{2+}-activated anion currents indicates that other factors may be important for activation of this conductance and suggests that Ca^{2+} may activate anion currents by interaction with intermediate components rather than directly binding to anion channels or that additional mechanisms may interact with anion channels.

DISCUSSION

We have demonstrated that elevation of the cytosolic-free Ca^{2+} concentration from the submicromolar to the micromolar concentration regulates two types of strongly voltage-dependent ion channels in the plasma membrane of *Vicia faba* guard cells: K^+ uptake channels and anion channels. Inhibition of K^+ uptake channels by cytosolic Ca^{2+} (12) can provide a molecular mechanism for physiologically observed Ca^{2+}-dependent inhibition of stomatal opening (2, 15) and Ca^{2+}-dependent inhibition of radioactively labeled K^+ (Rb^+) uptake (8).

Recent patch clamp studies of endosperm cells show that cytosolic Ca^{2+} activates cation releasing channels in the plasma membrane (16). We cannot exclude that elevation of cytosolic Ca^{2+} has an additional (third) effect on guard cell ion channels by direct modulation of K^+-releasing channels, as these channels were blocked in our experiments by perfusing the cytosol with minutely permeant Cs^+ ions (11, 12).

Ca^{2+} and voltage-dependent activation of anion channels in the plasma membrane of guard cells induces membrane depolarization (12). Therefore, anion channel-induced depolarization can activate voltage-dependent K^+-releasing channels (14). Simultaneous opening of anion and K^+ release channels can lead to a membrane potential intermediate to the equilibrium potentials of K^+ and anions, thus driving the physiologically observed, simultaneous release of K^+ and anions (7) through anion and K^+ channels. The Ca^{2+} activation of anion channels therefore provides a mechanism for the Ca^{2+} requisite during physiological induction of stomatal closing by abscisic acid (2, 9, 10, 15).

Recent single channel recordings with high chloride concentrations in the cytoplasm have shown single anion-selective channels with a voltage-dependence similar to that of the Ca^{2+}-activated anion currents described here (6). These voltage-dependent channels were also shown to be permeable to malate ions (6). This finding is of physiological significance as evidence suggests that guard cells release malate ions during stomatal closing (10, 17). The assumption that the steady-state voltage dependence of single Cl^- channels (6) and the time-dependent voltage dependence of the Ca^{2+}-activated anion conductance (12) are different from one another (6) appears to be questionable as the steady-state voltage dependences of anion channels and anion currents are similar. Further studies will be required to test the suggestion of Keller *et al.* (6) that guard cells require two different types of anion channels.

The Ca^{2+}-regulated anion conductance is activated by depolarizing the membrane potential to values more positive than approximately -80 mV (Fig. 1B, Fig. 2B; 6, 12). Therefore, mechanisms which can "predepolarize" the cell to potentials more positive than -80 mV may be necessary for the activation of the Ca^{2+} and voltage-dependent anion channels. Opening of plasma membrane Ca^{2+} channels as well as inhibition of H^+ pumps would be feasible candidates for predepolarization. Future patch clamp studies may provide insight into physiological mechanisms of predepolarization.

ACKNOWLEDGEMENTS

We thank Kelly Frawley for preparation of the manuscript. This research was supported by a grant from the NIH to S.H. and Alexander von Humboldt and NIH fellowships to J.I.S. We thank Dr. C. Schobert for reading the manuscript.

LITERATURE CITED

1. **Blatt MR** (1988) Potassium-dependent bipolar gating of K+ channels in guard cells. J Membrane Biol **102**: 235-246
2. **De Silva DLR, Cox RC, Hetherington AM, Mansfield TA** (1985) Synergism between calcium ions and abscisic acid in preventing stomatal opening. New Phytol **101**: 555-563
3. **Gotow K, Shimazaki K, Kondo N, Syono K** (1984) Photosynthesis-dependent volume regulation in guard cell protoplasts from *Vicia faba* L. Plant Cell Physiol **25**: 671-675
4. **Hamill OP, Marty A, Neher E, Sakmann B, Sigworth FJ** (1981) Improved patch-clamp techniques for high-resolution current recording from cells and cell-free membrane patches. Pflugers Arch Ges Physiol **391**: 85-100
5. **Hepler PK, Wayne RA** (1985) Calcium and plant development. Rev Plant Physiol **36**: 397-439.
6. **Keller BU, Hedrich R, Raschke K** (1989) Voltage-dependent anion channels in the plasma membrane of guard cells. Nature **341**: 450-453
7. **MacRobbie EAC** (1981) Effects of ABA in "isolated" guard cells of *Commulina communis* L. J Exp Bot **32**: 563-572
8. **MacRobbie EAC** (1986) Calcium effects in stomatal guard cells. In AJ Trewavas, D Marmé, eds, Molecular and Cellular Aspects of Ca^{2+} in Plant Development. Plenum, NY, pp 383-384
9. **Outlaw WH** (1983) Current concepts on the role of potassium in stomatal movements. Physiol Plant **59**: 302-311
10. **Raschke K** (1979) Movements of stomata. In W Haupe, ME Feinlieb, eds, Encyclopedia of Plant Physiology, Vol 7. Springer-Verlag, Berlin, pp 384-441
11. **Schroeder JI** (1988) K^+ transport properties of U^+ channels in the plasma membrane of *Vicia faba* guard cells. J Gen Physiol **92**: 667-683
12. **Schroeder JI, Hagiwara S** (1989) Cytosolic calcium regulates ion channels in the plasma membrane of *Vicia faba* guard cells. Nature **338**: 427-430
13. **Schroeder JI, Hedrich R, Fernandez JM** (1984) Potassium-selective single channels in guard cell protoplasts of *Vicia faba*. Nature **312**: 361-362

14. **Schroeder JI, Raschke K, Neher E** (1987) Voltage dependence of K^+ channels in guard-cell protoplasts. Proc Natl Acad Sci USA **84**: 4108-4112
15. **Schwartz A, Ilan N, Grantz DA** (1988) Calcium effects on stomatal movement in *Commulina communis* L. Plant Physiol **87**: 583-587.
16. **Stoeckel H, Takeda K** (1989) Calcium-activated, voltage-dependent, non-selective cation currents in endosperm plasma membrane from higher plants. Proc R Soc (London) **237**: 213-231
17. **Van Kirk CA, Raschke K** (1978) Release of malate from epidermal strips during stomatal closure. Plant Physiol **61**: 474-475

CALCIUM TRANSPORT IN THE UNICELLULAR GREEN ALGA *MESOTAENIUM CALDARIORUM*: A MODEL SYSTEM FOR PHYTOCHROME-MEDIATED RESPONSES

TOM BERKELMAN AND J. CLARK LAGARIAS

Department of Biochemistry and Biophysics, University of California, Davis, CA 95616, USA

Chloroplast orientation in the unicellular green alga *Mesotaenium caldariorum* and its filamentous relative, *Mougeotia* sp., is regulated by light (10). In both organisms, the single, plate-like chloroplast can be induced to move to optimize light harvesting under light-limiting conditions or to minimize light damage in the presence of excess light. This light-dependent chloroplast movement is red/far-red (R/FR) reversible, thus implicating phytochrome as the photoreceptor for the response (11).

Experimental evidence supports the involvement of the calcium ion as a second messenger in phytochrome responses (12, 19). The role of calcium in chloroplast rotation in *Mesotaenium* and *Mougeotia* is particularly well studied. Extracellular calcium is required for phytochrome-dependent chloroplast movement in *Mougeotia* (24). The uptake and release of calcium from *Mougeotia* and *Mesotaenium* cells and protoplasts have been shown to be R/FR reversible (6, 26). Chloroplast movement can be stimulated in the absence of light by application of the calcium ionophore A23187 to *Mougeotia* filaments (21). Calmodulin, which is present in both algal species (25, 13), has also been implicated to play a role in chloroplast movement. Calmodulin antagonists inhibit phytochrome-dependent chloroplast movement in *Mougeotia* (21, 25). Phytochrome-dependent chloroplast movement in *Mougeotia* is also inhibited by cytochalasin B, which suggests the involvement of actin filaments (23). This implies a role for calcium and calmodulin by analogy to actin-based movements in other systems.

Another notable feature of phytochrome-regulated chloroplast movement in both algal species is the action dichroism exhibited by the response. Using a linear polarized actinic light source, Haupt and colleagues have established that the phytochrome photoreceptor must be associated with a rigid structural component of the algal cell (10). Based on laser microbeam experiments, these investigators have ruled out an association of phytochrome with the chloroplast and have proposed that phytochrome is associated with the plasma membrane.

To date, virtually all of the studies of phytochrome signal transduction pathways in *Mesotaenium* and *Mougeotia* have used *in vivo* experimental approaches. In order to identify and characterize the molecular entities involved in the signal transduction pathway for chloroplast movement, we have initiated research to develop an experimental system to link phytochrome photoconversion to a biochemically measurable response *in vitro*. We have chosen *Mesotaenium caldariorum* as the experimental organism for our studies for several reasons. First, the organism is unicellular, therefore the complexity introduced by the presence of multiple cell types is avoided. Second, the organism is readily cultured in defined media. By contrast, *Mougeotia* is filamentous and is less easily grown in defined media. Third, recent studies in our laboratory have shown that *Mesotaenium caldariorum* is amenable to biochemical dissection, which is primarily due to the development of methods to prepare stable algal protoplasts in high yield (2, 14). Based on the arguments presented above, we have chosen calcium transport in this organism as a biochemical process which is potentially regulated by phytochrome. Specifically, we hope to address the following questions. Does phytochrome regulate any calcium transport process in *Mesotaenium caldariorum*? Are calcium-specific channels or calcium-transporting ATPases the molecular targets for phytochrome regulation? Which intracellular membranes are phytochrome-responsive? Does phytochrome modulate calcium transport in this organism directly? Or, does it do so indirectly, via interaction with a G-protein, a protein kinase, or a phospholipase which releases a second messenger? Figure 1 illustrates, in cartoon form, a number of hypothetical participants in the phytochrome-mediated signal transduction pathway.

In order to develop an *in vitro* phytochrome-responsive system from *Mesotaenium caldariorum*, two lines of investigation are presently under way in our laboratory. In one study, we have initiated experiments to purify phytochrome from *Mesotaenium* and recently have succeeded in isolating the intact algal photoreceptor (14). In the second investigation, we plan to use the purified photoreceptor and membrane fractions to reconstitute an effect of light on calcium transport systems from *Mesotaenium caldariorum*. This report summarizes our ongoing investigations on calcium transport in *Mesotaenium caldariorum* which is, in part, the subject of a submitted manuscript (2). Particular emphasis will be placed here on experiments designed to demonstrate the role of light as a regulator of calcium transport as measured *in vitro*.

METHODS AND MATERIALS

Inositol-(1,4,5)-trisphosphate (IP_3) was purchased from Calbiochem (San Diego, CA). Inositol-(1,4)-bisphosphate (PIP_2) and A23187 were purchased from Sigma (St. Louis, MO). Dibromo-BAPTA was purchased from Molecular

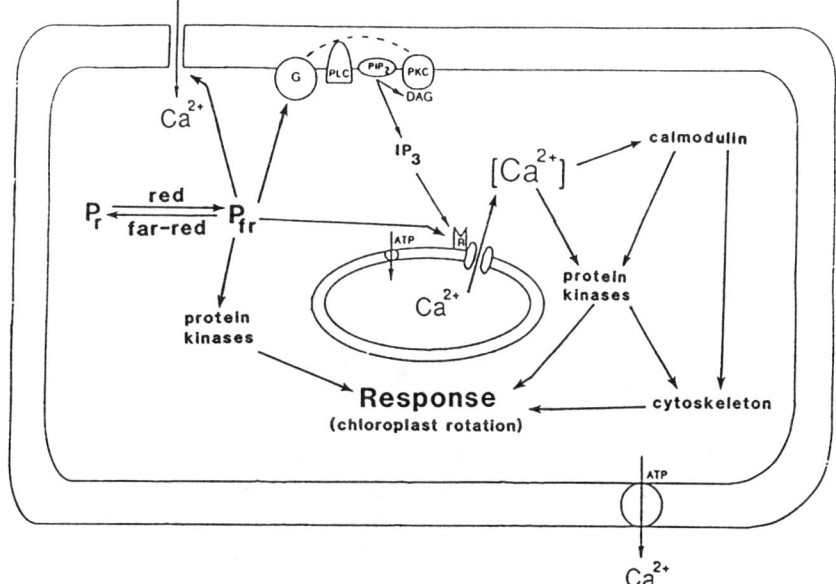

Figure 1. Possible elements in the signal transduction of the phytochrome response. The physiologically active form of phytochrome, Pfr, may exert its effect through a number of different mechanisms. It may interact directly with a calcium channel, either on the plasma membrane or on an internal membrane. It may also activate protein kinases or a G-protein (G). Interaction with a G-protein may activate phospholipase C (PLC), which catalysis the hydrolysis of PIP$_2$ to diacylglycerol (DAG) and IP$_3$. IP$_3$ would act in turn by opening a calcium channel on an internal membrane. Elevated cytosolic calcium leads to the response through the action of calcium-binding regulatory proteins such as calmodulin and calcium-dependent protein kinases, which cause cytoskeletal changes resulting in the movement response.

Probes (Eugene, OR). ^{45}CaCl$_2$ was purchased from Amersham (Arlington Heights, IL) or ICN (Costa Mesa, CA).

Cell growth, protoplast preparation, protoplast lysis, fractionation, calcium transport assays, marker analysis and the details of calcium electrode construction and use were as described (2).

Irradiations were performed using a 100-W incandescent bulb held 15 to 20 cm from the material to be irradiated. Red light was provided using a 660-nm interference filter. Far-red light was provided using 722- to 732-nm interference filter.

The K_ms for calcium were determined in the following assay medium: 250 mM sorbitol, 75 mM KCl, 25 mM Mes-Bis-Tris pH 6.2, 5 mM MgSO$_4$, 1 mM dithiothreitol, 1 mM ATP, 500 nM nigericin, and 100 µM CaCl$_2$ containing approximately 3.6 µCi/mL ^{45}Ca. Free Ca^{2+} in the medium was adjusted using varying concentrations of the calcium chelator dibromo-BAPTA. Dibromo-BAPTA was chosen as a calcium buffer for these experiments for a number of reasons. Its calcium affinity is in the appropriate range for these experiments. It has negligible affinity for Mg^{2+} ions. There is little effect of pH on its calcium affinity at physiologically relevant pHs and it is available in highly pure form (9, 22). The concentration of the calcium buffer was varied between 50 and 250 μM, and the free calcium concentration was calculated using information from Harrison and Bers (9) and Tsien (22). The membranes to be assayed were fractionated on a sucrose gradient and material corresponding to the peak of activity within the gradient was pooled. A portion of this fraction was diluted into the medium described above, and aliquots were removed at 2, 8, 16, and 24 min. Calcium uptake was determined by filtration (2).

Calcium uptake was determined in whole protoplasts as follows. Protoplasts of *Mesotaenium* were suspended in 600 mM sorbitol, 1 mM NaCl, 1 mM $NaHCO_3$, 200 μM KCl and 100 μM $CaCl_2$ to a density of 12.5 x 10^7 cells/mL. At time 0, $^{45}CaCl_2$ was added to approximately 0.13 μCi/mL. Aliquots of 400 μL were removed at various times and the protoplasts were separated from their medium by centrifugation through a layer of silicone oil. The aliquot containing the cells was pipetted into a microfuge tube containing (from the top down) a layer (100 μL) of 600 mM sorbitol, 50 mM $CaCl_2$, 1 mM NaCl, 1 mM $NaHCO_3$, and 200 μM KCl; a layer of silicone oil (Sigma melting point bath oil); and a layer of 600 mM sucrose. This was done so that the aliquot of protoplasts only mixed with the top layer. The tubes were spun in a microfuge for 15 s and frozen in liquid nitrogen. The tubes were cut through the silicone layer, and the lower portion of the tube containing the protoplasts in the 600 mM sucrose was placed in a vial containing scintillation fluid. Uptake of $^{45}Ca^{2+}$ into the protoplasts was determined by scintillation counting.

RESULTS AND DISCUSSION

Membranes exhibiting ATP-dependent calcium transport, as determined by $^{45}Ca^{2+}$ uptake, separate into two pools when subjected to centrifugation on 10 to 40% sucrose density gradients (Fig. 2). Some of the activity is associated with very low density membranes and sediments at the interface between the overlay solution and the 10% sucrose. The rest of the activity enters the gradient and is associated with vesicles with a median density of 1.11 g/mL. The distribution of the calcium transport activity most closely resembles that of two tonoplast markers, vanadate-insensitive ATPase, and pyrophosphatase (2). The distribution does not match that of markers for plasma membrane or mitochondria, which band sharply with higher median densities. Although markers for endoplasmic reticulum show more complex profiles, the major peaks of activity are also found at higher densities (Fig. 2).

Microsomes isolated from lysates of *Mesotaenium* protoplasts are an exceptionally good source of ATP-dependent calcium transport activity (2). Calcium transport can therefore be monitored with a calcium electrode which

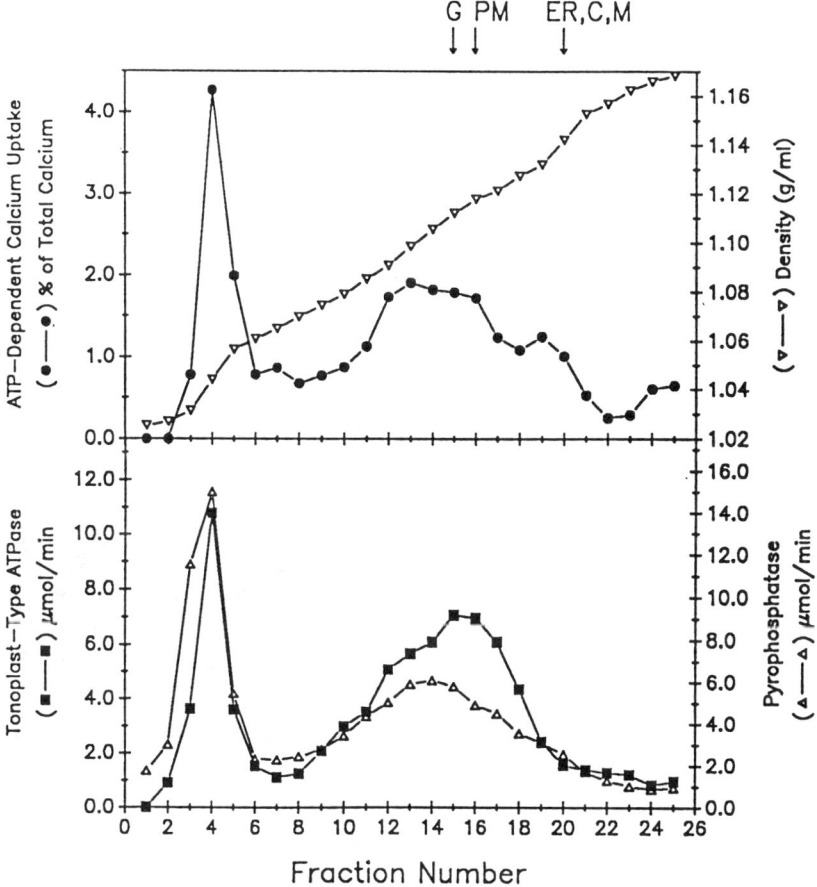

Figure 2. Distribution of ATP-dependent calcium uptake and tonoplast markers on a 10-40% sucrose gradient. The letters and arrows above the upper graph refer to the position of maximum activity for markers specific for various organelles. G, Golgi apparatus; PM, plasma membrane; ER, endoplasmic reticulum; C, chloroplast membranes (thylakoids); M, mitochondrial membranes.

allows sensitive real-time measurement of calcium concentration in the suspending medium. Microsomal suspensions containing less than 1 mg/mL of membrane protein are capable of lowering several hundred-fold the concentration of free calcium (Fig. 3). The free calcium concentration in these suspensions can be maintained stably at submicromolar levels in the presence of ATP. This corresponds to reported concentrations of cytoplasmic calcium in algal cells (18). Using such a system, we initiated experiments to effect the stimulated release of Ca^{2+} from the loaded vesicles. Microsomal suspensions have been used to study this phenomenon in both animal- and plant-derived systems. In many cases, this has been found to occur through the action of the

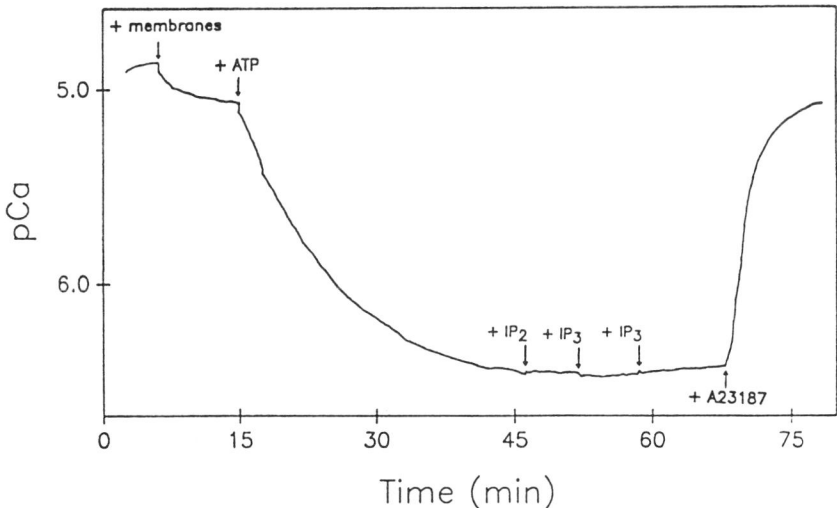

Figure 3. ATP-dependent uptake of calcium monitored with a calcium-selective electrode. The free calcium concentration in a stirred suspension of microsomes (360 µg/mL membrane protein) was monitored with a calcium-selective electrode. The calibrations on the ordinate were obtained using calcium buffer standards (pCa = -log[Ca^{2+}]). P_2 (10 µM), IP_3 (10 µM), and the calcium ionophore A23187 (5 µM) were added where indicated.

second messenger IP_3 (3, 7, 20). In some systems, however, this compound has been found to be inactive (17). We have investigated the effect of IP_3, IP_2, and light on the release of calcium from *Mesotaenium* microsomal vesicles. As shown in Figure 3, neither of these compounds had any detectable effect on the concentration of free calcium in the medium. Treatment of the microsome suspension with light was similarly ineffective (data not shown).

The inactivity of IP_3 in this system does not necessarily rule out the involvement of this second messenger in the modulation of intracellular calcium in *Mesotaenium*. Soluble factors not included in the system may be necessary for the effect. Components of the system may also become modified during preparation of the microsomes. Similarly, light-stimulated calcium release from microsomes might be possible in a reconstituted system containing the photoreceptor phytochrome, as well as other soluble components of the signal-transduction pathway.

External stimuli have also been shown to have an effect on the enzyme systems responsible for ATP-dependent calcium transport both in animals, where epinephrine stimulates calcium transport in cardiac sarcoplasmic reticulum (15), and in higher plants, where light treatment of the tissue affects the kinetic properties of calcium transport in isolated microsomes (4). We have therefore examined the possibility of light affecting the kinetics of calcium transport in *Mesotaenium* microsomes as well. Protoplasts were made from cells that were either placed in total darkness for 3 d (dark-adapted), or from control cells grown as described and harvested during the light period of the light-dark cycle. Prior to lysis, half of the protoplasts from dark-adapted cells were given 20 min of illumination with white light. The other half were kept under dim green light. The K_m for calcium in nigericin-insensitive, ATP-dependent calcium transport was determined in membranes isolated from the three groups of protoplasts (dark-adapted, dark-adapted with light treatment, and control) (Fig 4). The apparent K_ms determined were 5.0 μM, 6.2 μM and 4.5 μM, for dark-adapted, dark adapted plus light treatment and control cells, respectively. This preliminary experiment shows that any effect of light treatment on the K_m for calcium in isolated membranes is minor or non-existent.

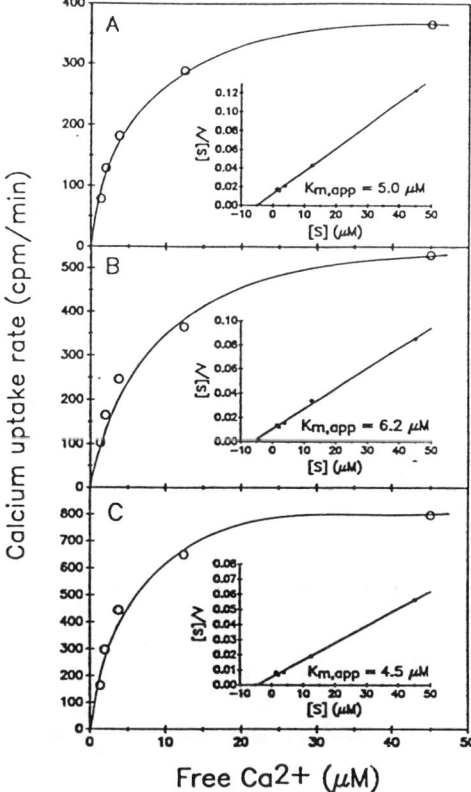

Figure 4. Determination of apparent K_ms for calcium transport. The calcium uptake rate was determined as described in Methods and Materials. Varying concentrations of free calcium were obtained using dibromo-BAPTA as a calcium buffer. Apparent K_ms were determined using Hanes-Woolf plots. A, Membranes from cells dark-adapted for 3 d. B, membranes from cells dark-adapted for 3 d and illuminated with white light for 20 min prior to lysis. C, Membranes from cells grown under the usual light/dark regime and harvested during the light period.

Lack of a light effect on the K_m of this particular partially-purified transport system does not rule out an effect of light on the kinetics of calcium transport *in vivo*. Many calcium transport systems are operative in cells, and the particular system assayed here may be irrelevant with respect to signaling. The K_ms determined (4.5-6.2 μM) are higher than those commonly reported for protonophore-insensitive calcium transport, and this activity may not be the one primarily responsible for maintaining submicromolar cytoplasmic calcium. Other calcium transport systems may have gone undetected in this study because the assay conditions were inappropriate, because the enzymes were inactivated during membrane isolation, because the membranes involved have been rendered permeable to calcium, or because the membranes under study may have the wrong orientation or "sidedness." A light effect on the kinetic parameters of this calcium transport system may also be overlooked because light-modulated regulatory factors are lost during the membrane isolation process.

The plasma membrane has previously been implicated in signal transduction of the phytochrome response. In higher plants, phytochrome has been found to be specifically associated with the plasma membrane (8, R. Hertel, personal communication). Phytochrome has been proposed to be associated with the plasma membrane of algal cells based on action dichroism experiments (10). It has also been shown that calcium uptake into cells of *Mesotaenium* and *Mougeotia* is phytochrome regulated (6, 26). Evidence from

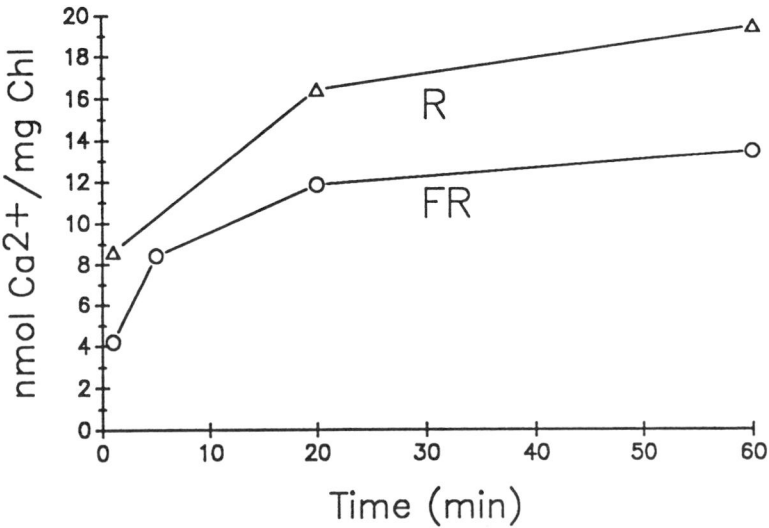

Figure 5. Uptake of $^{45}Ca^{2+}$ in whole protoplasts. Uptake of the tracer was determined at various times by centrifugation of the protoplasts through a layer of silicone oil. R, protoplasts under continuous illumination with red light; FR, protoplasts under continuous illumination with far-red light.

our work also supports this tenet. $^{45}Ca^{2+}$ added to the suspension medium of *Mesotaenium* protoplasts is taken up more rapidly and to a greater degree when the protoplasts are under red light illumination than when they are under far-red light illumination (Fig. 5). We have not, however, ruled out the possible involvement of photosynthesis in this effect. Because plasma membranes do not separate well from other membranes on sucrose density gradients, work is under way to purify *Mesotaenium* plasma membrane by other methods.

Although the experiments discussed have failed, to date, to demonstrate an effect of light or potential second messengers on calcium transport in *Mesotaenium* microsomes, the methodologies developed in this study will aid in the development of an *in vitro* system for the study of phytochrome function. The availability of purified *Mesotaenium* phytochrome will also be useful in the development of reconstituted systems (4). We intend to continue the investigation of active calcium transport in membranes purified from *Mesotaenium* cells, particularly the plasma membrane, in which calcium transport has already been shown to be light-regulated in an *in vitro* system derived from a higher plant (8). Recently developed methods for the production of plasma membrane vesicles of defined orientation should prove useful in these studies (16). Future experiments also include investigations of the possible regulation by phytochrome of plasma membrane calcium channels. These channels can be identified using drugs which have been shown to bind to the plasma membrane of higher plant and algal cells and block calcium uptake (1, 5). Channel characterization can be accomplished with experiments involving patch clamping, channel reconstitution, or calcium flux from inside-out plasma membrane vesicles.

ACKNOWLEDGMENTS

This work was supported by a grant (DCB87-04266) to J.C.L. from the National Science Foundation.

LITERATURE CITED

1. **Andrejauskaus E, Hertel R, Marme D** (1985) Specific binding of the calcium antagonist [^{3}H]verapamil to membrane fractions from plants. J Biol Chem 260: 5411-5414
2. **Berkelman T, Lagarias JC** (1990) Calcium transport in the green alga *Mesotaenium caldariorum*: preliminary characterization and subcellular distribution. Plant Physiol (in press)
3. **Dawson AP, Irvine RF** (1984) Inositol(1,4,5) trisphosphate-promoted Ca^{2+} release from microsomal fractions of rat liver. Biochem Biophys Res Commun 120: 858-864
4. **Dieter P, Marme D** (1983) The effect of calmodulin and far-red light on the kinetic properties of the mitochondrial and microsomal calcium-ion transport system from corn. Planta 159: 277-281
5. **Dolle R** (1988) Isolation of plasma membrane and binding of the Ca^{2+} antagonist nimodipine in *Chlamydomonas reinhardtii*. Physiol Plant 73: 7-14
6. **Dreyer EM, Weisenseel MH** (1979) Phytochrome mediated uptake of Calcium in *Mougeotia* Cells. Planta 146: 31-39

7. Drobak BK, Ferguson IB (1985) Release of Ca^{2+} from plant hypocotyl microsomes by inositol-1,4,5-trisphosphate. Biochem Biophys Res Commun 130: 1241-1246
8. Eisinger WR, Short TW, Briggs WR (1989) Light regulation of calcium fluxes in isolated membrane vesicles from pea (*Pisum sativum* L.) seedlings (Abstract). In European Symposium, Photomorphogenesis in Plants, Programme and Abstracts, #33
9. Harrison SM, Bers DM (1987) The effect of temperature and ionic strength on the apparent Ca-affinity of EGTA and the analogous Ca-chelators BAPTA and dibromo-BAPTA. Biochim Biophys Acta 925: 133-143
10. Haupt W (1982) Light-mediated movements of chloroplasts. Annu Rev Plant Physiol 33: 205-233
11. Haupt W, Thiele R (1961) Chloroplastenbewegung bei *Mesotaenium*. Planta 56: 388-401
12. Hepler PK, Wayne R (1985) Calcium and plant development. Annu Rev Plant Physiol 36: 397-439
13. Jacobshagen S, Altmueller D, Grolig F, Wagner G (1986) Calcium pools, calmodulin and light-regulated chloroplast movements in *Mougeotia* and *Mesotaenium*. In AJ Trewavas, ed, Molecular and Cellular Aspests of Calcium in Plant Development. Plenum Press, NY, pp 201-209
14. Kidd DG, Lagarias JC (1990) Phytochrome from the green alga *Mesotaenium caldariorum*. purification and preliminary characterization. J Biol Chem (in press)
15. Kirchberger MA, Tada M, Katz AM (1974) Adenosine 3':5'-monophosphate-dependent protein kinase-catalyzed phosphorylation reaction and its relationship to calcium transport in cardiac sarcoplasmic reticulum. J Biol Chem 249: 6166-6173
16. Larsson C, Askerlund P, Palmgren MG, Fredriksson K, Sommarin M, Widell S (1989) Sealed inside-out and right-side-out plasma membrane vesicles: - optimal conditions for formation and separation (Abstract). In European Symposium, Photomorphogenesis in Plants, Programme and Abstracts, #140
17. Lew RR, Briskin DP, Wyse RE (1986) Ca^{2+} uptake by endoplasmic reticulum from zucchini hypocotyls. The use of chlorotetracycline as a probe for Ca^{2+} uptake. Plant Physiol 82: 47-53
18. Okazaki Y, Yoshimoto Y, Hiramoto Y, Tazawa M (1987) Turgor regulation and cytoplasmic free Ca^{2+} in the alga *Lamprothamnium*. Protoplasma 140: 67-71
19. Roux SJ, Wayne RO, Datta N (1986) Role of calcium ions in phytochrome responses: an update. Physiol Plant 66: 344-348
20. Schumaker KS, Sze H (1987) Inositol-1,4,5-trisphosphate releases Ca^{2+} from vacuolar membrane vesicles of oat roots. J Biol Chem 262: 3944-3946
21. Serlin BS, Roux SJ (1984) Modulation of chloroplast movement in the green alga *Mougeotia* by the Ca^{2+} ionophore A23187 and by calmodulin antagonosts. Proc Natl Acad Sci USA 81: 6368-6372
22. Tsien RI (1980) New calcium indicators and buffers with high selectivity against magnesium and protons: design, synthesis, and properties of prototype structures. Biochemistry 19: 2396-2404
23. Wagner G, Haupt W, Laux A (1972) Reversible inhibition of chloroplast movement by cytochalasin B in the green alga *Mougeotia*. Science 176: 808-809
24. Wagner G, Klein K (1978) Differential effect of calcium movement in *Mougeotia*. Photochem Photobiol 27: 137-140
25. Wagner G, Valentin P, Dieter P, Marme D (1984) Identification of calmodulin in the green alga *Mougeotia* and its possible function in chloroplast reorientational movement. Planta 162: 62-67
26. Weisenseel MH (1986) Uptake and release of Ca^{2+} in the green algae *Mougeotia* and *Mesotaenium*. In AJ Trewavas, ed, Molecular and Cellular Aspects of Calcium in Plant Development. Plenum Press, NY, pp 193-199

STRUCTURAL AND FUNCTIONAL ANALYSIS OF HEAT SHOCK PROTEINS

H. MICHAEL HARRINGTON, STEFAN MOISYADI, AND YING TANG LU

Department of Plant Molecular Physiology, University of Hawaii, Manoa, Honolulu, HI 96822, USA

All biological organisms, including field-grown crops, undergo the heat shock response (HSR) when subjected to moderate upshift in growing temperature. The major features of the HSR in plants are identical to other systems; however, plants also synthesize a low molecular weight complex (LMWC) of heat shock proteins (HSPs) (16). All available evidence indicates that one or more of the HSPs are essential for the development and maintenance of thermotolerance (16).

Alterations in Membranes During Heat Stress

Inducers of the HSR stimulate phospholipid turnover in animal cell lines in a "G" protein-mediated response (4). Inositol trisphosphate (IP_3), diacylglycerol (DAG), and intercellular Ca^{2+} rapidly increased during heat shock (HS), suggesting the involvement of second messenger mechanisms. The breakdown of phosphatidylinositol 4,5 phosphate (PIP_2) to IP_3 and DAG may directly disrupt cell structure because PIP_2 appears to be involved in the anchoring of cytoskeletal microfilaments to the cell surface (12). The metabolism of PIP_2 from the plasma membrane during HS may affect membrane structure and function by altering lipid and lipid-protein interactions. This, in conjunction with the increased unsaturation of acyl chains observed during elevated temperature, may promote the formation of a hexagonal II (H_{II}) phase for lipids having two acyl chains (17). Plant cells contain all of the components of the phosphoinositol (PI) pathway (18), but there are no data on the effects of HS on IP_3 levels.

Small HSPs in maize (18 kD) (6) and soybean (15 kD) (13) are localized in the plasma membrane. Ultrastructural studies demonstrated that the soybean plasma membrane is disrupted during lethal HS (14) and that moderate HS resulted in the membrane becoming more permeable to solutes (13). Reduced solute leakage correlated with the association of the 15-kD HSP with the plasma membrane (13). These results suggest that HSPs may be involved in stabilizing membrane HS.

Calcium, Calmodulin, and Heat Shock

Several studies have examined the role of Ca^{2+} in stress responses. Rat hepatoma cells produced glucose stress-regulated proteins, but failed to produce HSPs when pretreated with EGTA (10). When the EGTA treatment occurred after HS, no inhibition of HSP synthesis was observed, suggesting that Ca^{2+} may be essential to initiate HSP synthesis. Alternatively, other workers indicate that Ca^{2+} is not essential in the HSR and HSP synthesis in *Drosophila* (7). The calmodulin (CaM) gene in *Chlamydomonas* contains a promoter motif similar to the HS element (22), suggesting that expression of this gene could be HS-regulated. Other workers have identified calmodulin-binding proteins (CaMBPs) as HSPs (21); however, the roles of Ca^{2+} and CaM in the HSR are currently unclear.

Little is known about the role of HSPs in the molecular mechanisms of thermotolerance. We have developed a comparative approach aimed at the functional identification of CaM-binding HSPs and key membrane-localized HSPs in sugarcane and tobacco cells.

METHODS

Tobacco (Wisconsin-38) and sugarcane cells (H50-7209) were cultured as suspensions; HS, radiolabeling of HSPs, and SDS-PAGE were as previously described (8, 15).

Tobacco Calmodulin-Binding Proteins

Tobacco cells were homogenized and the buffer extract applied to CaM sepharose-4B (5 mL) in the presence of 0.5 mM $CaCl_2$ (20). The column was step eluted with 20 column volume each of wash buffer (0.1 mM $CaCl_2$), wash buffer + 0.15 M NaCl, and wash buffer + 0.3 M NaCl. The bound fractions were eluted with a pulse of 1 mM EGTA. All fractions were dialyzed, lyophilized, and separated by SDS-PAGE.

Sugarcane Protein Kinase Activity

Microsomes and soluble fractions were isolated from control cells (23°C) or heat shock (36°C, 2 h) and phosphorylated in the presence of 55 mM Hepes, pH 7.4, 11 mM $MgCl_2$, 0.45 mM Na_2EGTA, 0.55 mM $CaCl_2$, 86 nM γ-^{32}P-ATP. Lamelli sample buffer (2X) was added, and reactions were terminated in a boiling water bath (10 min). After *in vitro* phosphorylation of native substrates or histone H_1 (3), labeled proteins were separated by SDS-PAGE.

RESULTS AND DISCUSSION

Tobacco Calmodulin-Binding Proteins

Preliminary experiments screening heat-shocked tobacco cells for the presence of CaMBPs using the [^{125}I]-CaM gel overlay method suggests that there are numerous CaMBPs in tobacco and that several are either HSPs or are enhanced by HS. Most notable of these are two 15- and 17-kD HSPs; however, neither of these exhibited Ca^{2+}-reversible binding of CaM (data not shown). While the overlay procedure is widely used, false-positives are detected for certain proteins, and the sensitivity of this procedure is insufficient to detect the minute quantities of these proteins expected to be synthesized during HS. A more direct approach using CaM-sepharose affinity chromatography was used to detect newly synthesized, labeled HSPs that bind to CaM. Autoradiographs indicate that several polypeptides from control and HS cells bind to CaM-sepharose in a Ca^{2+}-dependent manner even in the presence of 0.3 M NaCl (Fig. 1, lanes A7 and B7). Elution of the column with EGTA after 0.15 M NaCl washes produced similar results (lanes A8 and B8). The salt washes release selected sets of HSPs from the column, and some of these may also be CaM targets. The HSPs that exhibit Ca^{2+}-dependent binding are not labeled under control conditions and do not bind to sepharose-4B (Lu and Harrington, unpublished data). Putative CaMBPs/HSPs range in apparent mol wt from approximately 111 to 15 kD. Of these, HSPs of 68, 22, and 21 kD exhibit binding of biotinylated CaM on nitrocellulose blots. We

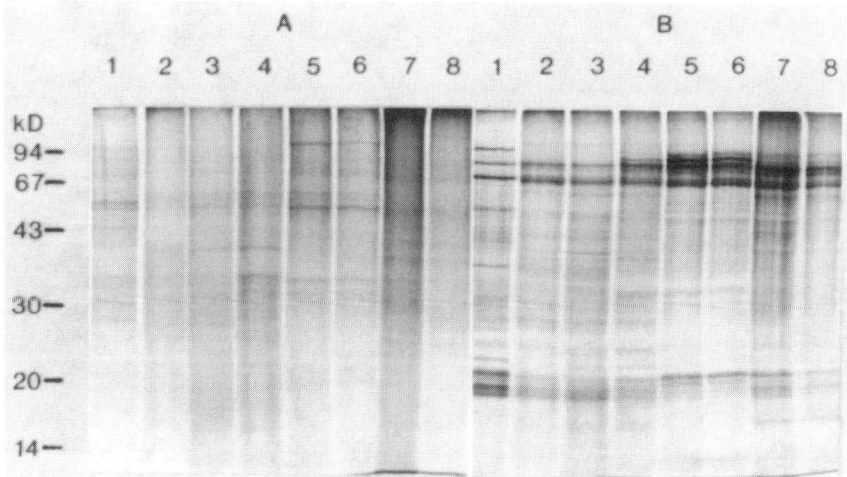

Figure 1. Calmodulin sepharose affinity chromatography of tobacco proteins. Autoradiogram of labeled proteins. Panel A, control cells; panel B, HS cells. Lane 1, total Lamelli SDS sample buffer extract; lane 2, total column buffer extract; lane 3, 0.5 mM CaCl$_2$ column wash; lane 4, 0.1 mM CaCl$_2$ column wash; lane 5, 0.15 M NaCl column wash; lane 6, 0.3 M NaCl column wash; lane 7, 1 mM EGTA column wash; lane 8, 5 mM EGTA column wash after 0.15 M NaCl.

are presently confirming the ability of these fractions to bind CaM in overlay studies using ^{125}I-CaM and ^{35}S-CaM, as well as determining the subcellular localization of CaMBPs.

The identity of the CaMBPs/HSPs in tobacco, and their functions and relationship to the HSR, are presently unknown; however, possible roles include a CaM-dependent protein kinase or other enzyme that may act to alter existing metabolic processes during HS. Alternatively, a HSP may also act to sequester CaM, limiting disassembly of microtubules and preventing additional disruption of the cytoskeleton during HS.

Sugarcane Protein Kinase Activity

A detailed characterization of the sugarcane HSR and of HSP synthesis (15) and other unpublished subcellular localization studies focused our attention on 18-kD membrane HSPs whose synthesis and turnover strongly correlate with thermotolerance.

Figure 2 shows time courses of proteins phosphorylated *in vitro* for purified microsome and the soluble post-microsomal supernatant fractions obtained from control and HS cells. These results demonstrate time-dependent

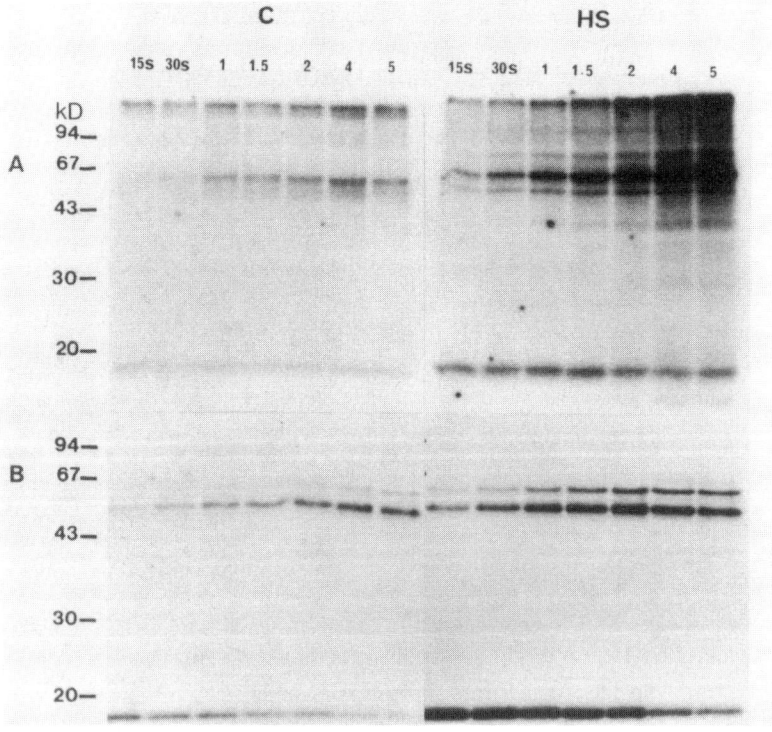

Figure 2. *In vitro* phosphorylation of sugarcane extracts. Autoradiogram of labeled proteins. Panel A, microsomes; panel B, post-microsomal soluble supernatant, incubation times from 15 s to 5 min.

increases in ^{32}P incorporation into peptides of 40 kD and larger. The overall levels of phosphorylation are increased 10- to 20-fold in the HS samples. These results could be explained if HS induced *de novo* synthesis of phosphorylatable protein substrates, by increased protein kinase activity, or a combination of both. There is a major phosphorylated 18-kD peptide that exhibits an anomalous labeling pattern--that is, high levels of label are present in the 18-kD peptide(s) at very short times (15 s), followed by progressive losses as the time course proceeds. This behavior is exemplified by the pattern seen in the HS supernatant fraction. Interestingly, the supernatant fraction contains numerous polypeptides as visualized on Coomassie stained gels, but only peptides in the 50- and 60-kD range are phosphorylated. Both of these peptides are also heavily represented in the microsome fractions. These experiments were carried out at 0°C in an ice bath, but similar, more rapid results, were obtained in 23°C incubations. The phosphorylation patterns for the sugarcane 18-kD protein(s) are similar to those reported for an 18-kD catalytic subunit of an autophosphorylating protein kinase (APK) found in pea plasma membrane. The possibility that HS induces an 18-kD APK subunit is supported by the fact that the 18-kD HSPs and the putative APK subunit activity are not induced at 32°C, but both are sharply enhanced at 36°C. It is unlikely that this result is due to the presence of a HS-induced kinase substrate because *in situ* phosphorylation of proteins on gels after SDS-PAGE demonstrate HS-stimulated autophosphorylation of 18-kD peptides, even in the presence of large amounts of histone H_1 in the gel matrix. (Moisyadi and Harrington, unpublished results). Under these conditions, the amount of phosphorylation activity would be limited by the amount of enzyme rather than substrate. Proteins in the 18-kD range are the only peptides labeled *in situ*.

The major sugarcane 18-kD HSP and the putative 18-kD APK subunit copurify by selective solubilization of microsomes followed by ion exchange chromatography. Active APK fractions contain only HSPs in the 18-kD range. The active APK fraction contains a major and a minor phosphorylatable peptide with mobilities corresponding respectively to a major and a minor ^{35}S methionine-labeled 18-kD HSP in 2-dimensional PAGE. The major peptide exhibits pronounced ^{32}P label turnover as would be expected of the APK subunit as described by Trewavas and co-workers (1, 3). In addition, the active APK fraction catalyzes time-dependent phosphorylation of histone H_1 *in vitro*. Thus, our data suggest that a major 18-kD HSP in sugarcane is analogous to the catalytic subunit of an APK. It is unlikely that this protein is involved in the initial steps of thermotolerance because of the temporal regulation of the LMWC. However, Trewavas and co-workers have demonstrated that the pea APK (pp18) phosphorylates PI to form phosphoinositol phosphate (PIP) (2). Thus, if the IP_3 pathway is involved in the plant cell HSR, the APK may play a role in restabilizing the membrane and the cytoskeleton.

CONCLUSION

Transmembrane signaling pathways often utilize plasma membrane-associated protein kinases (9) and/or the phosphatidylinositol pathway (4). Members of the protein kinase family are regulators of a variety of essential cellular functions including gene activation, protein synthesis, energy metabolism, and cytoskeleton assembly/stability (9). One of the most common events during HS is elevated protein phosphorylation. A primary example is the apparent phosphorylation of the HSF, which coincides with the activation of HS genes (19). The well-documented roles of the IP$_3$ pathway, the Ca^{2+}/CaM system, and protein phosphorylation status in regulatory mechanisms suggest that the HS-induced CaMBPs and APK play important roles in the HSR.

LITERATURE CITED

1. **Blowers DP, Trewavas AJ** (1987) Autophosphorylation of plasma membrane bound calcium dependent protein kinase from pea seedlings and modification of catalytic activity by autophosphorylation. Biochem Biophys Res Commun **143**: 691-696
2. **Blowers D, Trewavas AJ** (1988) Phosphatidylinositol kinase activity of plasma membrane-associated calcium-activated protein kinase from pea. FEBS Lett **238**: 87-89
3. **Blowers DP, Trewavas AJ** (1989) Rapid cycling of autophosphorylation of a Ca^{2+}-calmodulin regulated plasma membrane located protein kinase from pea. Plant Physiol **90**: 1279-1285
4. **Calderwood SK, Stevenson MA, Hahn GM** (1987) Heat stress stimulates inositol trisphosphate release and phosphorylation of phosphoinositides in CHO and Balb C 3T3 cells. J Cell Physiol **130**: 369-376
6. **Cooper P, Ho THD** (1987) Intracellular localization of heat shock proteins in maize. Plant Physiol **84**: 1197-1203
7. **Drummond IAS, McClure SA, Poenie M, Tsien RY, Steinhardt RA** (1986) Large changes in intracellular pH and calcium observed during heat shock are not responsible for the induction of heat shock proteins in *Drosophila melanogaster*. Mol Cell Biol **6**: 1767-1775
8. **Harrington HM, Alm DM** (1988) Interaction of heat and salt shock in cultured tobacco cells. Plant Physiol **88**: 618-625
9. **Hunter T** (1987) A thousand and one protein kinases. Cell **50**: 823-829
10. **Lamarche S, Chretien P, Landry J** (1985) Inhibition of the heat shock response and synthesis of glucose-regulated proteins in Ca^{2+}-deprived rat hepatoma cells. Biochem Biophys Res Commun **131**: 868-876
11. **Landry J, Chretien P, Lambert H, Hickey E, Weber LA** (1989) Heat shock resistance conferred by expression of the human HSP27 gene in rodent cells. J Cell Biol **109**: 7-15
12. **Lassing I, Lindberg U** (1985) Specific interaction between phosphatidylinositol 4,5-biphosphate and profilactin. Nature **314**: 472-474
13. **Lin CY, Chen YM, Key JL** (1985) Solute leakage in soybean seedlings under various heat shock regimes. Plant Cell Physiol **26**: 1493-1498

14. **Mansfield AM, Lingle WL, Key JL** (1988) The effects of lethal heat shock on nonadapted and thermotolerant root cells of *Glycine max*. J Ult Mol Struct Res **99**: 96-105
15. **Moisyadi S, Harrington HM** (1989) Characterization of the heat shock response in cultured sugarcane cells. Plant Physiol **90**: 1156-1162
16. **Nagao RT, Key JL** (1989) Heat shock proteins of plants. Cell Cult Som Cell Gen Plants **6**: 297-328
17. **Rilfors L, Lindblom G, Wieslander A, Christiansson A** (1984) Lipid bilayer stability in biological membranes. *In* M Kates, LA Manson, eds, Biomembranes, Vol 12. Plenum Press, NY, pp 205-245
18. **Sommarin M, Sandelius AS** (1988) Phosphatidylinositol and phosphatidylinositolphosphate kinases in plant plasma membranes. Biochim Biophys Acta **958**: 268-278
19. **Sorger PK, Pelham HRB** (1988) Yeast heat shock factor is an essential DNA-binding protein that exhibits temperature dependent phosphorylation. Cell **54**: 855-864
20. **Watterson DM, Vanaman TC** (1976) Affinity chromatography purification of a cyclic nucleotide phosphodiesterase using immobilized modulator protein, a troponin C-like protein from brain. Biochem Biophys Res Commun **73**: 40-46
21. **Widada JS, Ferraz C, Asselin J, Trave G, Colote S, Haiech J, Marti J, Liautard JP** (1989) Cloning and deletion mutagenesis using direct protein-protein interaction on an expression vector. Identification of the calmodulin binding of α-fodrin. J Mol Biol **205**: 455-458
22. **Zimmer WE, Schloss JA, Silflow CD, Youngblom J, Watterson DM** (1988) Structural organization, DNA sequence, and expression of the calmodulin gene. J Biol Chem **262**: 19370-19383

INTERACTION OF CALCIUM AND AUXIN IN THE REGULATION OF ROOT ELONGATION

MICHAEL L. EVANS, HELEN G. KISS,
AND HIDEO ISHIKAWA

Department of Botany, The Ohio State University, Columbus, OH 43210, USA

It is well established that calcium (Ca) is important for plant growth and development (2, 9). Recently, evidence has accumulated linking calcium and growth hormone action in plants (10, 12). It has been proposed that plant hormone effects on growth and development are mediated by hormone-induced shifts in cytoplasmic Ca levels, with Ca acting as a secondary messenger to bring about the ultimate effects of the hormone. We have examined this idea using auxin inhibition of maize root elongation as a model system. In particular, we have: *i*) compared the kinetics of auxin-induced and Ca-induced inhibition of root growth, *ii*) tested the concentration dependence of Ca inhibition of root growth, *iii*) examined the Ca dependence of auxin action in roots, and *iv*) tested the effects of applied Ca on cytoplasmic Ca levels in maize root protoplasts.

MATERIALS AND METHODS

Plant Material

Seedlings of maize (*Zea mays* L., either hybrid B73 x Missouri 17 or cv Merit) were raised as described previously (10). Seedlings were used when the roots were 2 to 2.5 cm long (about 3 d after planting).

Growth Measurements

Root elongation was measured using the transducer method described by Evans (6). Measurement of localized surface extension rates was done with a modified version of the video digitizer system described previously (14). Small dots of Speedball oil-base paint were placed along the root surface and the velocity of movement of each mark relative to the root tip was determined from slopes of position versus time curves. Localized relative growth rates (% h^{-1}) were determined from the slopes of the velocity/position curves. Roots were allowed to grow in distilled water for 100 min prior to any experimental treatments.

Measurement of Cytoplasmic Calcium Levels

Cytoplasmic Ca levels were measured in protoplasts prepared from the elongation zone of maize roots. Protoplasts were prepared using a modification of the methods of Lin (13). Protoplasts were loaded with the potassium salt of the fluorescent Ca indicator dye, fura-2, by incubating in 20 μM fura-2-K^+ at pH 4.5 for 1 to 2 h (see ref. 3). Estimates of cytoplasmic Ca concentration in protoplast suspensions were made by measuring the 340/380 nm excitation ratio using a Perkin-Elmer LS-5B luminescence spectrometer according to the methods of Grynkiewicz *et al.* (8).

RESULTS

Concentration Dependence of Calcium Inhibition of Maize Root Elongation

Figure 1 shows results from the studies of Hasenstein and Evans (10) on the concentration dependence of Ca inhibition of maize root elongation. Calcium caused transient growth inhibition. The inhibition began almost immediately, and both the magnitude and duration of the inhibition increased with increasing concentration of Ca applied. Even with 5 mM Ca, the growth rate returned to a value equal to or slightly greater than the rate prior to treatment within about 80 min. Hasenstein and Evans found the kinetics of

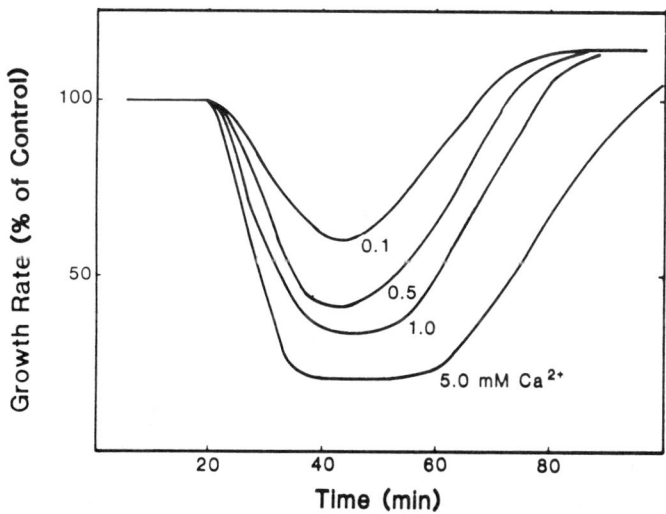

Figure 1. Concentration dependence of the effect of Ca (as $CaCl_2$) on the growth of primary roots of maize seedlings (hybrid B73 x Missouri 17). Growth medium = 5 mM Mes buffer, pH 6.2. Figure reproduced from Hasenstein and Evans (10) with permission from American Society of Plant Physiologists.

growth inhibition by 1 mM Ca was nearly identical to inhibition by 0.1 μM IAA except that inhibition by Ca was immediate, while inhibition by IAA occurred after a latent period of about 15 min (Fig. 2). They speculated that auxin inhibition of root growth may be mediated by Ca, with auxin acting somehow to elevate cytoplasmic Ca levels at least transiently. This idea was supported by the observation that auxin inhibition was greatly diminished in roots depleted of Ca (see Fig. 4 in ref. 10). We have tested this idea further by comparing the profiles of auxin and Ca inhibition in maize roots as described in the next section.

Comparative Patterns of Localized Growth Rate Profiles in Roots Treated with Auxin or Calcium

Figure 3 shows profiles of relative growth rate *versus* position in a maize root growing in distilled water before and after addition of 0.2 μM IAA. In the absence of IAA, there was a broad growth zone extending from about 1.5 to 7.5 mm behind the root tip (peak = about 4 mm). IAA suppressed growth uniformly throughout the elongation zone (see 30 min curve in Fig. 3). After about 50 min, the growth rate began to recover, even though the root remained in IAA. At 70 min after the addition of IAA, growth in the basal part of the elongation zone (about 4.5-8 mm) remained severely inhibited, while growth in the apical region had recovered. Surprisingly, the growth rate within the apical part of the elongation zone following recovery from IAA inhibition was substantially greater than that prior to addition of auxin, *i.e.* auxin promoted the growth of these cells after a period of transient inhibition.

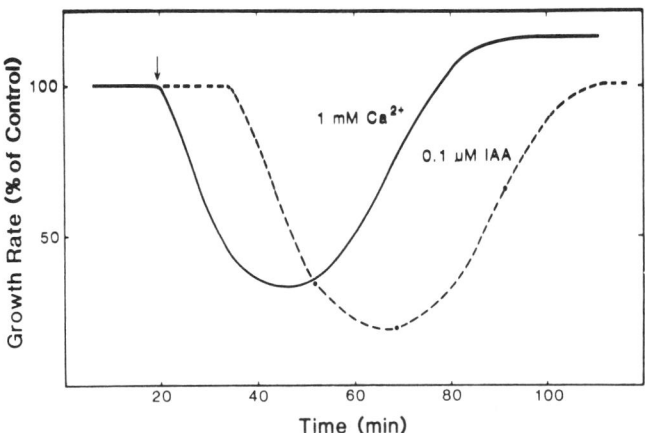

Figure 2. Comparative action of IAA and Ca (as CaCl$_2$) on the growth of primary roots of maize seedlings (hybrid B73 x Missouri 17). Growth medium as in Figure 1. Figure reproduced from Hasenstein and Evans (10) with permission from American Society of Plant Physiologists.

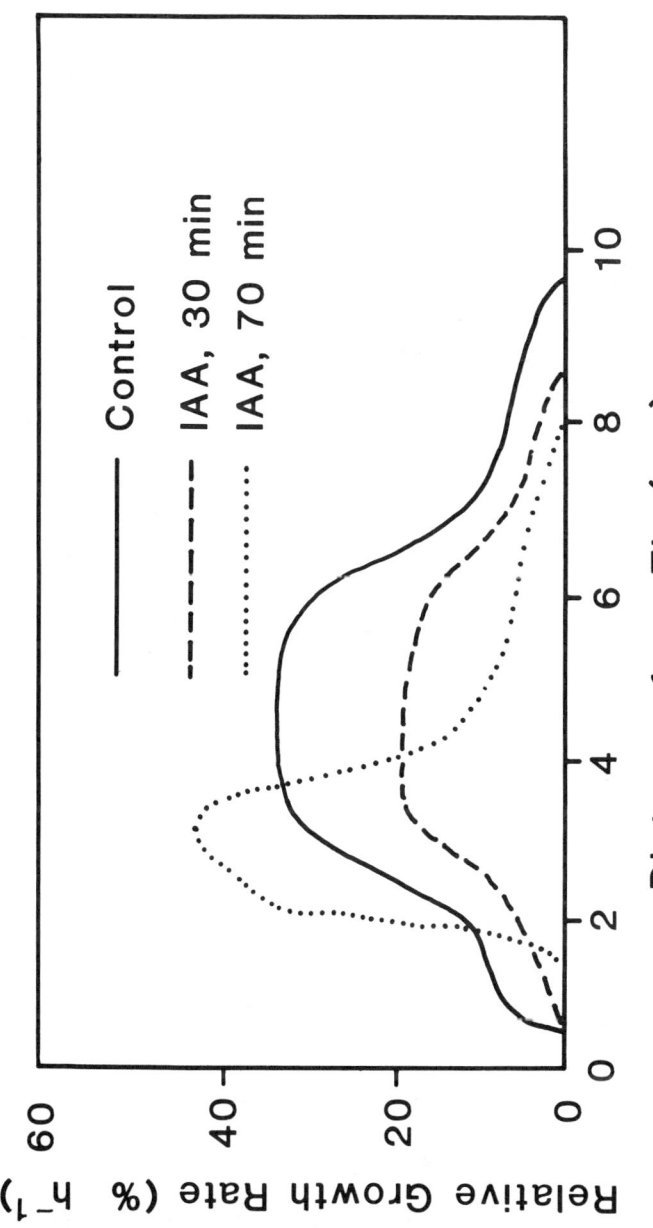

Figure 3. Relative growth rate as a function of position in primary roots of maize treated with 0.2 μM IAA. Growth medium = distilled water.

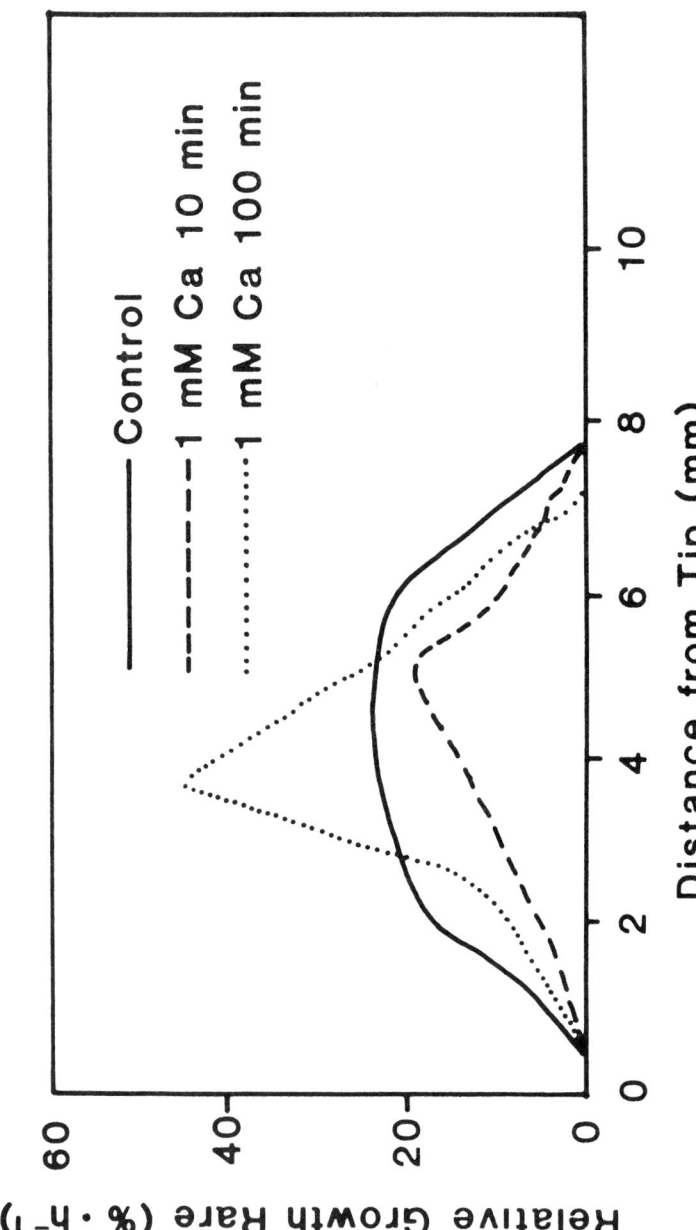

Figure 4. Relative growth rate as a function of position in primary roots of maize treated with 1 mM $CaCl_2$. Growth medium = distilled water.

When roots were treated with 1 mM Ca, growth was suppressed throughout the elongation zone, although the inhibition in the region of peak elongation was less severe (Fig. 4). Following inhibition by Ca, the growth rate of the root began to recover. However, unlike the case with IAA, recovery occurred primarily in cells in the center of the elongation zone where growth following recovery from Ca inhibition was much greater than that prior to addition of Ca, *i.e.* Ca promoted the growth of these cells after a period of transient inhibition.

Effect of Applied Calcium and Auxin on Cytoplasmic Calcium Levels

Figure 5 shows data from measurements of intracellular Ca levels in protoplasts prepared from the elongation zone of maize roots. Fluorescence of protoplasts loaded with 20 μM fura-2-K$^+$ was strong throughout the cytoplasm, but weak or absent in the vacuole (data not shown). Upon addition of 2 mM calcium chloride, there was no apparent change in fluorescence, indicating no change in intracellular-free Ca levels. Subsequent addition of (3-[cholamidopropyl)-dimethylammonio]-1-propanesulfonate (CHAPS) to the medium led to enhanced fluorescence indicating elevation of cytoplasmic Ca. Addition of manganese chloride (5) reduced fluorescence strongly when added after CHAPS (Fig. 5), but had no effect when added prior to treatment with CHAPS (data not shown).

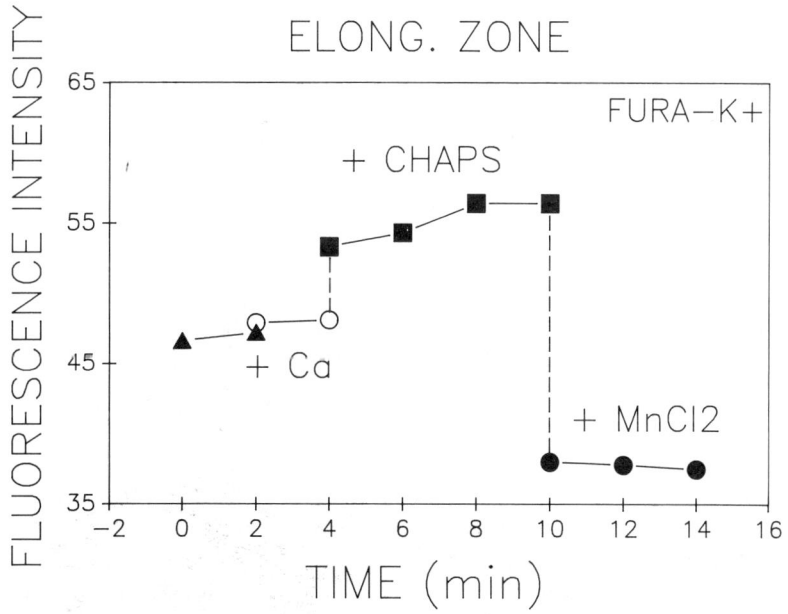

Figure 5. Relative fluorescence intensity (340 nm) of elongation zone protoplasts preloaded with fura-2. Protoplasts were treated sequentially with 2 mM CaCl$_2$, 10% CHAPS, and 10 mM MnCl$_2$ as indicated.

DISCUSSION

Both Ca and auxin cause rapid, though transient, inhibition of root elongation in maize seedlings. Also, roots of maize depleted of Ca show reduced sensitivity to growth inhibition by auxin (10). These observations are consistent with the hypothesis (10) that auxin-induced shifts in cytoplasmic Ca mediate the inhibitory action of auxin on root growth.

More detailed comparisons of localized growth profiles in maize roots, however, reveal clear differences between the pattern of growth inhibition and recovery following treatment with auxin *versus* Ca. Upon treatment of roots with 0.2 μM IAA, growth is inhibited throughout the elongation zone. However, cells in the basal half of the elongation zone remain inhibited, while cells in the apical portion begin to grow again and, in fact, develop a rate of elongation greater than that displayed prior to auxin treatment. This accounts for the partial recovery from growth inhibition in roots treated with IAA. Similar data on auxin-induced growth inhibition and recovery have been reported for roots of wheat (1, 11) and *Phleum* (7).

In roots treated with 1 mM Ca, growth is also inhibited throughout the elongation zone, just as with IAA. However, the pattern of growth recovery in the presence of Ca is different from that during recovery from auxin inhibition. In the presence of Ca, following the initial inhibition, growth remains inhibited in both the basal and apical regions of the elongation zone, but recovers strongly in the central portion of the elongation zone. In roots recovering from Ca inhibition, the growth rate of cells in the center of the elongation zone greatly exceeds that occurring prior to Ca treatment. These data indicate that, following an initial period of inhibition, IAA promotes the growth of cells in the apical region of the elongation zone, while Ca promotes the growth of cells in the central region of the elongation zone. Thus, it seems unlikely that long-term auxin action on the growth of maize roots is mediated simply by auxin-induced changes in intracellular Ca levels. However, since the initial phase of inhibition is similar following addition of Ca or IAA, we cannot rule out the possibility that the initial inhibitory action of auxin on root growth is mediated by changes in intracellular Ca levels.

Direct addition of Ca to protoplasts from the elongation zone of maize roots did not cause a change in the intracellular concentration of Ca. This indicates that the effects of Ca on root growth may be mediated by Ca action at an extracellular site. However, this conclusion assumes that: *i*) the permeability of protoplasts to applied Ca is comparable to, or at least not less than, that of intact cells, and *ii*) that the putative shifts in intracellular Ca required to affect growth are large enough to be detected using fura-2. Taken at face value, our data indicate that Ca action on growth is brought about by changes in apoplastic Ca levels. Cleland (4) has suggested that the inhibition of coleoptile elongation by Ca is mediated by a Ca-induced increase in the pH of the Donnan free space, thus causing reduced acid-mediated wall loosening and, hence, reduced growth. Our data are consistent with this hypothesis.

ACKNOWLEDGMENTS

This research was supported by National Science Foundation grant DMB8608673 and by National Aeronautics and Space Administration grant NAGW-297. We thank Greg Hatosky for assistance in raising the seedlings.

LITERATURE CITED

1. Burström H (1957) On the adaptation of roots to β-indolylacetic acid. Physiol Plant **10**: 187-197
2. Burström HG (1968) Calcium and plant growth. Biol Rev **43**: 287-316
3. Bush DS, Jones RL (1987) Measurement of cytoplasmic calcium in aleurone protoplasts using indo-1 and fura-2. Cell Calcium **8**: 455-472
4. Cleland RE (1990) Calcium, cell walls and growth. *In* RT Leonard, PK Hepler, eds, Calcium in Plant Growth and Development. American Society of Plant Physiologists Symposium Series, Vol II, Rockville, MD, pp 9-16 (this volume)
5. Deber CM, Tom-Kun J, Mack E, Grinstein S (1985) Bromo-A23187: A nonfluorescent calcium ionophore for use with fluorescent probes. Anal Biochem **146**: 349-352
6. Evans ML (1976) A new sensitive root auxanometer. Preliminary studies of the interaction of auxin and acid pH in the regulation of intact root elongation. Plant Physiol **58**: 599-601
7. Goodwin RH (1972) Studies on roots. V. Effects of indoleacetic acid on the standard root growth pattern of *Phleum pratense*. Bot Gaz **133**: 224-229
8. Grynkiewicz G, Poenie M, Tsien RY (1985) A new generation of Ca indicators with greatly improved fluorescence properties. J Biol Chem **260**: 3440-3450
9. Hanson JB (1984) The functions of calcium in plant nutrition. *In* PB Tinker, A Läuchli, eds, Advances in Plant Nutrition, Vol 1. Praeger Publishers, NY, pp 149-208
10. Hasenstein KH, Evans ML (1986) Calcium dependence of rapid auxin action in maize roots. Plant Physiol **81**: 439-443
11. Hejnowicz Z (1961) The response of the different parts of the cell elongation zone in root to external β-indolylacetic acid. Acta Soc Bot Poloniae **30**: 25-42
12. Hepler PK, Wayne RO (1985) Calcium and plant development. Annu Rev Plant Physiol **36**: 397-439
13. Lin W (1980) Corn root protoplasts. Isolation and general characterization of ion transport. Plant Physiol **66**: 550-554
14. Nelson AJ, Evans ML (1986) Analysis of growth patterns during gravitropic curvature in roots of *Zea mays* by use of a computer-based video digitizer. J Plant Growth Regul **5**: 73-89

A CALCIUM- AND VOLTAGE-ACTIVATED K$^+$ CURRENT PROVIDES A MECHANISM TO RAPIDLY RESTORE THE MEMBRANE POTENTIAL FOLLOWING STIMULUS-INDUCED DEPOLARIZATION

Karen A. Ketchum and Ronald J. Poole

*Department of Biology, McGill University,
Montréal, Québec, H3A 1B1, Canada*

Transient depolarization of the cell membrane potential is associated with signal transduction pathways initiated by both hormonal and environmental stimuli. The molecular mechanisms which affect these changes are likely to include the coordinated activity of plasma membrane ion channels. We have previously described three types of K$^+$-dependent currents in protoplasts from corn suspension cells using the whole-cell recording technique (1). Here, we present evidence that one of these currents, the K$^+$ current evoked by membrane depolarization ($I_{K^+ out}$), is also Ca^{2+} dependent. This current is a plausible candidate for repolarizing the membrane potential following signaling events which incorporate membrane depolarization and changes in the cytosolic-free Ca^{2+} concentration.

To ascertain whether the activation of $I_{K^+ out}$ was dependent upon the influx of extracellular Ca^{2+}, compounds that antagonize plasma membrane Ca^{2+} transport were added to the external solution. $I_{K^+ out}$ was relatively insensitive to the ionic Ca^{2+}-channel blockers, Cd^{2+}, Co^{2+}, Mn^{2+}, and La^{3+}, exhibiting inhibition only when the external concentration of these ions exceeded 10 mM. In contrast, organic Ca^{2+}-channel blockers from the dihydropyridine (nitrendipine) and phenylalkylamine (verapamil, D600) families caused marked inhibition. The IC$_{50}$ for nitrendipine (concentration which produced 50% reduction in current) was 1 μM at a test potential of +60 mV with a 20-min incubation period. One possible interpretation of these data is that the activation of this K$^+$ current is dependent upon a rise in intracellular Ca^{2+} and that the pathway for Ca^{2+} transport is dihydropyridine and phenylalkylamine sensitive.

To further test this premise, we assayed $I_{K^+ out}$ in cells where either the calcium-buffering capacity, or the free Ca^{2+} concentration of the intracellular medium (pipette-filling solution), had been changed. When the buffering capacity of the intracellular medium was elevated by increasing the EGTA concentration from 4 to 40 mM, there was a corresponding reduction of $I_{K^+ out}$. At a test potential of +60 mV, the average decrease in conductance was 72% (n = 4). Conversely, raising the free Ca^{2+} concentration from 40 to 300 nM, with a constant EGTA level, enhanced the outward K^+ current, resulting in a two-fold increase in conductance. In addition, the voltage-activation range for I_{K^+out} was shifted in the higher Ca^{2+} medium toward more negative potentials.

Taken together, these results provide strong evidence for Ca^{2+}- and voltage-activated K^+ channels at the plasma membrane of corn. Furthermore, these channels appear to work in concert with a Ca^{2+} transport system which has pharmacological similarities (*i.e.* sensitivity to dihydropyridines and phenylalkylamines) to the L-type voltage-dependent Ca^{2+} channel found in animal tissues.

LITERATURE CITED

1. Ketchum *et al.* (1989) Plant Physiol **89**: 1184

IONIC REQUIREMENTS FOR POLARIZATION OF *PELVETIA* ZYGOTES

DARRYL L. KROPF AND SUSAN R. HURST

Department of Biology, University of Utah, Salt Lake City, UT 84112, USA

The fucoid zygote establishes an embryonic (rhizoid/thallus) axis in response to any number of external vectors, and is a paradigm for cell polarization during plant development. The earliest detectable expression of the axis is a transcellular ionic current which flows through the cell from rhizoid to thallus poles. A fraction of this current is carried by Ca^{2+}, and this Ca^{2+} circulation is thought to play a pivotal role in axis establishment. To date, the bulk of the current has not been characterized. Our aim is to understand more clearly the importance of specific ionic fluxes in early development. We have begun by conducting a series of experiments aimed at identifying those ions which are critical to the polarization process.

We have assayed populations of synchronously developing zygotes for their ability to photopolarize in a variety of ionic and osmotic milieus. When zygotes are exposed to unilateral light, they form an axis along the light vector and subsequently grow a rhizoid from the shaded hemisphere. In our experiments, zygotes were grown for 8 h in complete artificial seawater (ASW), under uniform lighting, and then transferred to the test medium and placed in unilateral light for 4 h. At the end of this period, they were transferred back into ASW and allowed to germinate in the dark. The position of rhizoid outgrowth was scored in relation to the orienting light pulse.

In the first set of experiments, the concentration of each ion in ASW was varied independently. ASW contains 0.45 M NaCl, 30 mM $MgCl_2$, 16 mM $MgSO_4$, 10 mM KCl, 9 mM $CaCl_2$, and is buffered to pH 8.5 with 10 mM Tris. Using buffers, the pH was adjusted over the range 4.5 to 9.7. Higher pHs could not be attained due to precipitation of salts from the medium. Polarization was unaffected between pH 5.5 and 9.7; at pH 4.5 most of the cells died. The potassium concentration was varied by simple deletion or addition; concentrations from 0 (no KCl added) to 100 mM KCl supported polarization equally well. To investigate sodium, NaCl was replaced with N-methylglucamine. Reducing Na^+ from 450 to 1 mM had no effect; below 1 mM, there was a slight reduction in polarization. The Ca^{2+} concentration was varied from 9 mM to 10 pM; concentrations below 25 μM were obtained using EGTA buffers. As was previously found in *Fucus*, polarization was not responsive to the Ca^{2+} concentration. Magnesium and sulfate were studied together by deletion and, again, no effects were seen. Finally, chloride removal (or substitution with sulfate) also did not block polarization. Taken together, these data indicate that polarization is not dependent upon the transcellular circulation of any one ion in the medium.

In a second set of experiments, salts were added back one at a time to a sucrose solution under conditions of constant osmolality. The sucrose solution alone did not support polarization on its own. Potassium salts (KCl or K_2SO_4) were sufficient for polarization, as was NaCl. The percent polarization increased linearly with increasing salt concentration up to 10 mM. Higher NaCl concentrations were toxic. Magnesium salts were less effective. Interestingly, calcium salts actually reduced polarization below that of sucrose alone. Calcium salts also inhibited the ability of 10 mM KCl to support polarization. We interpret this data to mean that the zygotes must generate sufficient turgor pressure in order to establish an axis. As yet, we do not know the meaning of the calcium results. The effects of varying turgor pressure were investigated by adding sucrose to ASW; above 0.6 M sucrose, there was a marked inhibition of polarization. We suggest that ion uptake in early development serves to increase turgor pressure and that the bulk of the transcellular ion current may reflect the spatial segregation of channels transporting osmotically active cations (K^+ and/or Na^+) from those taking up anions (Cl^- and/or SO_4^{2-}).

STUDIES OF CALMODULIN AND IONS IN THE RHIZOBIUM-LEGUME SYMBIOSIS

DAVID W. EHRHARDT, MELANIE J. BARNETT, AND SHARON R. LONG

*Department of Biological Sciences, Stanford University,
Stanford, CA 94305-5020, USA*

We are studying the symbiotic association of *Rhizobium meliloti* and its host, alfalfa (*Medicago sativa*), which results in formation of nitrogen-fixing nodules on the plant roots. In early stages of the association, the action of *Rhizobium* nodulation (*nod*) causes the single-celled root hairs which usually grow straight to grow in distorted or curled shapes by *Rhizobium*, and causes cells in the inner cortex of the plant root to show dedifferentiation and division, some layers away from the bacteria which are still on the outside of the root (2, 5). Analysis of both bacterial and plant mutant phenotypes suggests that the events in the root hair and in the inner cortex--proximal and distant effects of bacterial *nod*-gene action--may be mechanistically coupled (3, 5). One possible way in which this could happen would be the initiation in the root hair of a signal, which could be transduced into a secondary signal that stimulates cell divisions deeper within the root; alternatively, a single signal from the bacteria could cause the different effects in the root hairs and inner cortex cells if receptors for the signal were coupled to different cytoplasmic response systems (6). To understand the cell biology of signals and responses in the plant during early stages of *Rhizobium* infection, we are analyzing changes in membrane potential in root hair cells after bacterial inoculation, and changes in the production and/or localization of calmodulin.

To begin the study of calmodulin, we have constructed a cDNA clone bank from alfalfa seedling poly-A^+ RNA in a Lambda-Zap vector (Stratagene); this was screened with the cDNA clone of *Arabidopsis* calmodulin (1). Six clones were obtained in the original calmodulin screening. The sequence of one has been determined completely. It encodes an open-reading frame of 149 amino acids; it shows one amino acid difference in the N-terminal portion of the protein in comparison to the barley calmodulin sequence (4). We plan to determine the number of distinct calmodulin genes in the alfalfa genome and to study timing and localization of calmodulin expression in roots and in *Rhizobium*-provoked nodules.

To understand better the events occurring in root hairs during growth and in response to *Rhizobium*, we are developing methods of injecting into and measuring transmembrane potentials in intact root hairs. We have developed a standard preparation of an alfalfa seedling in a perfusion chamber, and have used micropipettes pulled to 0.2-micron outside diameter to penetrate the hair cell wall and measure transmembrane potentials. In a standard NOD-III medium (relevant ionic concentrations, 5 mM K^+, 2 mM Ca^{++}, 0.1 mM Na^+, 0.1 mM Cl^-, 0.5 mM^{++}, pH 6.0), the root hairs maintain a steady resting potential of about 125 mV for at least 1 h. We see a reproducible depolarization of about 10 to 15 mV following inoculation of root hairs with *Rhizobium*. We wish to determine whether this is related to nodulation. The micropipette apparatus will also be used for injecting calcium-dependent dyes.

LITERATURE CITED

1. Braam J, Davis RW (1990) Cell 60: 357-364
2. Dudley ME, Jacobs TW, Long SR (1987) Planta 171: 289-301
3. Dudley ME, Long SR (1989) The Plant Cell 1: 65-72
4. Ling V, Zielinski RE (1989) Plant Physiol 90: 714-719
5. Long SR (1989) Cell 56: 203-214
6. Long SR, Cooper J (1988) *In* Molecular Genetics of Plant-Microbe Interactions. APS Press, St. Paul, MN, pp 163-178

TRACHEARY ELEMENT DIFFERENTIATION INVOLVES AT LEAST TWO CALCIUM-REGULATED EVENTS—ONE CALMODULIN-DEPENDENT AND ONE CALMODULIN-INDEPENDENT

ALISON W. ROBERTS AND CANDACE H. HAIGLER

Texas Tech University, Department of Biological Sciences, Lubbock, TX 79409, USA

Suspension cultures of *Zinnia elegans* L. mesophyll cells can be induced to differentiate into tracheary elements (TEs) by the addition of cytokinin in the presence of appropriate concentrations of auxin (1). Differentiation is synchronous with characteristic secondary wall thickenings becoming evident at about 50 h after the addition of cytokinin; TEs are completely formed by about 80 h. This culture system provides an excellent model for studying the role of calcium and calmodulin in TE differentiation and, particularly, for investigating which events in this complex process are calcium- and calmodulin-regulated. Tracheary element differentiation includes the following sequence of events: response to cytokinin, gene activation, pattern formation, rearrangement of the cytoskeleton, deposition of secondary cell wall polysaccharides, lignification, and autolysis. Studies on other plants and animals indicate that these types of events are likely to be regulated by calcium and calmodulin. Furthermore, a rise in chlorotetracycline fluorescence just prior to the appearance of secondary wall thickenings in Zinnia cells indicates that calcium is involved in TE differentiation (2). We have now used calcium-channel blockers and calmodulin inhibitors, added and removed at different times during differentiation, in an attempt to unravel the roles of calcium and calmodulin in this complex process.

Calcium-channel blockers, including nifedipine, -202-791, and lanthanum, inhibited TE differentiation when added at 48 h (all times refer to the time elapsed from the initiation of culture), just prior to the appearance of secondary cell wall thickenings, indicating that calcium uptake is required for secondary wall deposition. This is supported by the complete reversibility of channel blockers added at 24 and 47.5 h, when they were removed at 48 h. When channel blockers were added at 0 h and removed at 48 h, inhibition was only partially reversed, indicating a second calcium-dependent event near the initiation of culture. Calmodulin antagonists inhibited TE differentiation when added at 0 to 24 h, but not when added after 24 h. Inhibition of differentiation by calmodulin inhibitors present during the first 24 h was reversed by washing in inductive medium with a high+ concentration of cytokinin, but not by washing in noninductive medium. This suggests that calmodulin is required for the response to cytokinin that initiates differentiation. Preliminary experiments with inhibitors of the phosphatidylinositol transduction pathway, lithium and neomycin, indicate that the phosphatidylinositol pathway may also play a role in the differentiation of TEs.

Our results are consistent with a model in which: (*i*) calmodulin regulates the response to cytokinin and possibly cytokinin-stimulated gene expression, (*ii*) rearrangement of the cytoskeleton is independent of calmodulin, and (*iii*) secondary cell wall deposition is dependent on calcium transport, but not calmodulin.

LITERATURE CITED

1. Fukuda H, Komamine A (1980) Plant Physiol **65**: 57-60
2. Roberts AW, Haigler CH (1980) Protoplasma (in press)

THE ROLE OF CALCIUM IN TOMATO ROOT GROWTH AND RESISTANCE TO *PHYTOPHTHORA PARASITICA* UNDER SALT STRESS

S. S. SNAPP AND C. SHENNAN

Vegetable Crops Department, University of California, Davis, CA 95616, USA

Adequate calcium levels are essential to promote healthy root growth under saline conditions. In particular, a favorable calcium-to-sodium ratio has been found to be crucial to K^+/Na^+ ion selectivity at the plasmalemma. As well, supplemental calcium protects root cell division and cell elongation under salinity stress, possibly acting at the cytoskeleton level. Concomitantly, phytopathologists have found that the amendment of soil with calcium will decrease infection of avocado and eucalyptus by Phytophthora species. The mechanism by which calcium enhances resistance remains unknown, although high levels of pectate-associated calcium in the middle lamelle have been implicated as an important deterrent to fungal invasion of epidermal tissue. Further, recent evidence demonstrates that resistance to *Phytophthora cinammomi* in a wide variety of plant species is related to genotypic ability to regenerate roots upon invasion by the fungus. As calcium can enhance root growth rates, one might expect addition of calcium to promote root regeneration and thus resistance to *Phytophthora* root rot. This effect would be most pronounced under saline conditions. Taken together, the literature indicated to us that calcium levels might regulate tomato response to salt, and thus have a strong influence on plant resistance to salt-enhanced Phytophthora root rot.

As a model system, we are interested in the interrelationship of calcium nutrition, salt stress, and the susceptibility of tomato to *Phytophthora parasitica*. We conducted a preliminary study under field conditions to investigate the influence of two levels of salinity and a variable sodium/calcium ratio on tomato root and shoot growth, calcium uptake and distribution, and severity of Phytophthora root rot. Soil salinity levels (EC_c) were established at approximately 1.1 ds/m (no added salt), 2.8 ds/m (two medium salt treatments at a 4:1 and a 24:1 $Na^+:Ca^{2+}$ ratio) and 4.6 ds/m (high salt). Increased severity of Phytophthora root rot in tomato cultivar UC82B was highly correlated with increasing salinity. By contrast, resistance to *P. parasitica* in cultivar CX8303 held up under all salt regimes. An enhanced calcium treatment at the moderate salt level (4:1 compared to 24:1 $Na^+:Ca^{2+}$ ratio) may have modified disease severity and improved yields in UC82B and

CX8303, but only to a limited extent. From preliminary data, calcium uptake and distribution differ markedly in the two cultivars.

Root growth was dramatically affected by salinity in UC82B, the susceptible cultivar, but not in the resistant cultivar, CX8303. Root regeneration may thus play an important role in tomato resistance to *P. parasitica*, as has been suggested for *P. cinammomi*. We hypothesize that supplemental calcium could protect root growth by improving ion selectivity of the plasmalemma for potassium over sodium, preventing excess sodium from being accumulated and improving the potassium supply to the root meristem. Further, ensuring adequate calcium nutrition under salt stress may improve cell division and elongation rates in the root, thus enhancing plant water status and ion uptake. Future research will focus on investigating the distribution of active roots in soil, root turnover rates, and plant nutrition (calcium, sodium potassium, and chloride levels) of UC82B and CX8303 as affected by sodium:calcium ratios and infection by *P. parasitica*. Spatial heterogeneity in the field may have influenced plant response to calcium in our preliminary study and further research will be conducted under homogeneous soil conditions.

CALCIUM-DEPENDENT PROTEIN KINASES FROM BARLEY

L. J. KLIMCZAK AND G. HIND

Biology Department, Brookhaven National Laboratory, Upton, NY 11973, USA

We are interested in the functional role of calcium-dependent protein kinases in signal transduction in plants. To provide a basis for such studies, we began a systematic biochemical characterization of these activities in barley. The initial investigation comprised partial purification of soluble and membrane-bound, calcium-dependent protein kinase activities and comparative analysis of their properties. Both forms showed an active polypeptide of 37 kD on activity gels with incorporated histone as substrate. They eluted from chromatofocusing columns at an identical isoelectric point of pH 4.25 ± 0.2, and also co-migrated on several other chromatographic affinity media including Matrex Gel Blue A, histone-agarose, phenyl-Sepharose, and heparin-agarose. Both activities co-migrated with chicken ovalbumin during gel filtration through Sephacryl S-200, indicating a native molecular mass of 45 kD. The activities share a number of enzymatic properties including salt and pH dependence, free calcium stimulation profile, substrate specificity, and K_m values. The soluble activity was shown to bind to artificial lipid vesicles. These data strongly suggest that the soluble and membrane-bound calcium-dependent protein kinases from barley are very closely related or even identical.

CELL- AND TISSUE-SPECIFIC PRECIPITATION OF CALCIUM OXALATE IN SEEDS

MARY ALICE WEBB

Department of Botany and Plant Pathology, Purdue University, West Lafayette, IN 47907, USA

Calcium is often sequestered in plant vacuoles as crystalline calcium oxalate (1, 2). Calcium oxalate deposits are very abundant in seeds, where they have characteristic patterns of distribution. The most common site of calcium oxalate deposition is in the seed coat. In many developing seeds, a cell layer within the integuments differentiates as a "crystal layer" in which each cell forms a crystal of calcium oxalate (3). Crystal layers in a variety of seeds have been examined by scanning electron microscopy to observe the structure and anatomical distribution of crystals within the mature seed coat (4). Some examples of prominent crystal layers will be illustrated in seed coats of *Carica papaya*, *Sesamum indicum*, and *Phaseolus vulgaris*. We have hypothesized that calcium oxalate deposition structurally reinforces the capacity of the seed coat to protect soft internal seed organs, in a manner analogous to the protection afforded by the shells of molluscs. In some taxonomic groups, deposits of calcium carbonate or silica are found in a similar distribution pattern.

Calcium oxalate is also deposited in the embryo in anatomically specific patterns. For example, in the embryo of *Cercis canadensis*, crystals form only in cells of the cotyledonary epidermis. Crystals were identified as calcium oxalate by a histochemical staining procedure, acid solubility tests, and x-ray microanalysis. Each cell of the epidermis contains a spherical aggregate of calcium oxalate, called a druse; druses are typically larger on the adaxial side than the abaxial side. Within the epidermal cells, druses are found inside storage protein bodies.

In certain taxonmic groups, calcium oxalate is very abundant in the endosperm. In several species of *Vitis*, approximately half the endosperm cells contain druses within storage protein bodies. These druses have a complex structure, consisting of an organic core, a region surrounding the core composed of both organic and crystalline material, and a peripheral region composed of interpenetrant crystals (5). Crystal composition was identified by powder x-ray diffraction as calcium oxalate monohydrate. Isolated developing druses were observed with scanning electron microscopy, and druses were examined *in situ* by transmission electron microscopy. As is typically the case in plants (see 1, 2), crystals form in association with an organic matrix that develops within the cell vacuole. Preliminary studies have been initiated to gain information about the biochemical characteristics of such matrices.

LITERATURE CITED

1. Arnott HJ, Pautard FGE (1970) Calcification in plants. *In* H Schraer, ed, Biological Calcification. Appleton-Century-Crofts, New York, pp 375-446
2. Franceschi VR, Horner HT Jr (1980) Bot Rev 46: 361-428
3. Netolitsky F (1926) Anatomie der Angiospermen-Samen. Gebrtuder Borntrager, Berlin
4. Webb MA, Arnott HJ (1982) Scanning Elect Micr 3: 1109-1131
5. Webb MA, Arnott HJ (1983) Scanning Elect Micr 4: 1759-1770

ISOLATION OF CLONED cDNA AND GENOMIC DNA SEQUENCES ENCODING CALMODULIN AND CALMODULIN-LIKE PROTEINS FROM BARLEY AND *ARABIDOPSIS*

RAYMOND E. ZIELINSKI, VINCENT LING, AND IMARA PERERA

Department of Plant Biology, University of Illinois, Urbana, IL 61801, USA

We have used cloned cDNA sequences encoding calmodulin (CaM) from barley (1) to isolate the corresponding barley genomic DNA sequences. The barley CaM gene contains no introns within the coding region, nor within the 3'-untranslated region. The promoter region of the gene, however, has not yet been isolated. Thus, we do not yet have information on *cis*-acting sequences that regulate CaM gene transcription. We have also isolated barley genomic sequences encoding a protein (BCaM-2) that shares considerable amino acid sequence similarity with barley CaM, which represents either a CaM isoform or a more distantly related member of the Ca-modulated protein family. The predicted amino acid sequence differences between barley CaM and BCaM-2 (for the region encoding residues 67-148), in general, represent conservative changes, many of which are observed in CaM proteins isolated from other sources. In contrast to the barley CaM gene, BCaM-2 contains an apparent intron within the fourth Ca^{2+}-binding domain.

The barley CaM cDNA sequences have also been used as probes to isolate cDNAs encoding two CaM isoforms and a 21.6-kD, CaM-like protein (p21) from an *Arabidopsis thaliana* cDNA library. Within the coding regions, the CaM isoform cDNAs differ by 15% in their nucleotide sequence, and of those nucleotide differences, 80% represent changes in codon third positions. At the amino acid sequence level, only four changes are predicted between the two *Arabidopsis* CaM isoforms: all of the changes are conservative and occur outside the Ca^{2+}-binding domains. The 3'-untranslated regions of the CaM cDNAs share little DNA sequence homology, clearly indicating that these molecules were derived from mRNAs transcribed from different genes. Northern blot hybridizations are being carried out using gene-specific CaM probes made from the 3'-untranslated regions of the two CaM cDNAs to assess the relative levels of expression of their respective mRNA sequences. The derived amino acid sequence encoding p21 shares 65% similarity with the CaM sequences, but only three alter proposed Ca^{2+}-binding ligands (N for D and H for C substitutions in loop 1, and a D for N substitution in loop 2). However, p21 has a number of unique structural attributes compared with known CaM proteins: there are numerous nonconservative amino acid substitutions throughout the molecule; an apparent three amino acid residue deletion in the region of the molecule between Ca^{2+}-binding loops 2 and 3; and a 42-amino acid C-terminal extension that shares no significant homology with any known Ca^{2+}-binding protein. Ca^{2+}-binding by p21, which was produced by *in vitro* translation in a wheat germ extract, was demonstrated by Ca^{2+}-induced mobility shifting during SDS-PAGE. Taken together, our results strongly indicate that there are likely to be numerous genes encoding Ca-modulated proteins in higher plants.

LITERATURE CITED

1. Ling V, Zielinski RE (1989) Plant Physiol **90**: 714-719

INTRACELLULAR-FREE CA^{2+} IN BARLEY PROTOPLASTS: INDO-1 FLOW CYTOMETRY

A. F. WRONA, J. R. AIST, J. P. SLATTERY, AND R. M. SPANSWICK

Boyce Thompson Institute (A.F.W.), *Department of Plant Pathology* (A.F.W, J.R.A.), *Biotechnology Institute* (J.P.S.), *and Section of Plant Biology* (R.M.S.), *Cornell University, Ithaca, NY 14853, USA*

Breeding lines of barley, *Hordeum vulgare* L., differing in their susceptibility to the powdery mildew fungus, *Erysiphe graminis* f. sp. *hordei*, form papillae (parasite-elicited wall appositions on the inner surfaces of the cell walls of host plants) in response to attempted infection by the fungus. The resistant barley isoline more rapidly forms larger papillae than does the susceptible isoline, and thereby successfully blocks penetration of the encroaching pathogen. Since papilla formation involves Ca^{2+}-mediated secretion, is abolished by the chelation of Ca^{2+}, and is inhibited by low exogenous Ca^{2+} and treatments with auxin, we postulated that an elevation of cytoplasmic-free Ca^{2+} plays a role in preventing powdery mildew in these breeding lines. The resistant isoline could have a constitutively higher level of cytoplasmic-free Ca^{2+} than the susceptible isoline, or there could be a transient rise in cytoplasmic-free Ca^{2+} in the host as a response to fungal attack. We used indo-1 and flow cytometry to measure cytoplasmic-free Ca^{2+} in protoplasts isolated from both isolines to determine if there is a constitutive difference that could help explain enhanced papilla formation in the resistant isoline.

A major advantage of flow cytometry is that in a few minutes, we were able to measure the responses of 10,000 indo-loaded protoplasts, as opposed to the relatively few measurements possible with imaging and photon counting methods. Fluorescence from each protoplast, excited by an argon laser tuned to 360 nm, was measured simultaneously at both emission wavelengths (405 and 495 nm), so we eliminated the problems of registration associated with indo imaging. Because protoplasts were illuminated for only 1 to 10 μs and measured only once as they flowed rapidly across the laser beam, we avoided measuring photobleached cells. Furthermore, the sensitivity of the photomultipliers and the greater intensity of indo-1 fluorescence greatly reduced both the amount of dye needed and the associated problems of phytotoxicity. Measurements and manipulations of the two isolines will be discussed.

Supported by USDA CRGO grants #85-CRCR-1-1669 and #87-CRCR-1-2306.

THE EFFECT OF CA^{2+} ON CHLOROPHYLL ACCUMULATION IN CUCUMBER (*CUCUMIS SATIVUS*) COTYLEDONS

Carol Réiss and Randy Wayne

Section of Plant Biology, Cornell University, Ithaca, NY 14853, USA

The first step in the greening of etiolated cucumber cotyledons involves the absorption of light by phytochrome. In order to test whether Ca^{2+} contributes to the signal transduction chain involved in red-light-stimulated chlorophyll accumulation, we treated etiolated cucumber cotyledons with Ca^{2+}-EGTA-buffered solutions and exposed them to 2 h of white light, ± a 2-min red-light pretreatment. The red-light-stimulated accumulation of chlorophyll is dependent on the concentration of external Ca^{2+}; 3 μM supports a half maximal response. The calcium-channel blockers nifedipine and NdCl₃ inhibit red-light-stimulated chlorophyll production, while the Ca^{2+}-selective ionophore A23187 stimulates synthesis of chlorophyll a and b. These data indicate that calcium does contribute to the signal transduction chain involved in red-light-stimulated chlorophyll synthesis. To determine if this effect is on the production of chlorophyll a/b-binding protein, native green gels are being used to determine levels of this chlorophyll-stabilizing protein under various Ca^{2+} states.

ULTRASTRUCTURAL LOCALIZATION OF FREE CALCIUM IN THE ORGANIC ACID AND CALCIUM-SECRETING TRICHOMES OF CHICKPEA (*CICER ARIETINUM* L.)

MARK D. LAZZARO AND WILLIAM W. THOMSON

Department of Botany and Plant Sciences, University of California, Riverside, CA 92521, USA

Multicellular trichomes present on the leaves, stems, and pods of chickpea (*Cicer arietinum* L.) secrete organic acids, primarily malate, and the pH of secretions is 1.0. The trichomes have 18 cells, including one basal cell, three elongate stalk cells, and 14 head cells arranged in four tiers (1). Using x-ray microanalysis, we found that calcium was also present in secretions. Using potassium pyroantimonate, which forms an electron-dense precipitate with free calcium, we found that free calcium was present in small amounts in both the apoplast and symplast throughout the trichome. In particular, we found calcium in the cytoplasm, vacuole, and ER of the stalk cells, and precipitates were quite dense in the plasmodesmata between the stalk cells. In the head cells, we found small amounts of calcium in the cytoplasm, mitochondria, ER, and nucleus. Additionally, the precipitate was quite dense in the membrane network along the periphery of the head cells. This membrane network may be a labyrinth of plasma membrane invaginations, and the presence of precipitate suggests that the network may be involved in calcium secretion.

LITERATURE CITED

1. Lazzaro MD, Thomson WW (1989) Can J Bot **67**: 2669-2677

Ca^{2+}-INDUCED SPACING PATTERNS OF Ca-OXALATE CRYSTALS IN DEVELOPING HICKORY LEAVES

ROLF BORCHERT

Department of Physiology and Cell Biology, University of Kansas, Lawrence, KS 66045-2106, USA

Leaves of all broad-leaved trees contain numerous Ca-oxalate crystals, usually arranged in characteristic spatial patterns. Analysis of developing leaves of *Carya ovata* (shagbark hickory) revealed that (*i*) Ca^{2+} entering leaves with the transpiration stream induces the differentiation of crystal cells, which precipitate Ca^{2+} as oxalate; and (*ii*) characteristic spacing patterns of crystal cells in leaves arise from the interaction between apoplastic Ca^{2+} and crystal cells acting as Ca^{2+}-sinks.

RESULTS AND CONCLUSIONS

In sections of immature hickory leaves floating on 2- to 4-mM Ca-acetate, crystal cells were induced within 6 h at densities much higher than those occurring during normal leaf development. Mesophyll cells of immature leaves thus can be induced by millimolar [Ca^{2+}] to transdifferentiate into crystal cells.

During the 400-fold enlargement of developing hickory leaves, a characteristic sequence of crystal spacing patterns is observed. Consecutive crystal patterns were quantified by computerized image-analysis, and the mechanism of pattern formation was deduced. Very small Ca-oxalate crystals appear in the mesophyll at high density (450 crystals mm^{-2}) shortly after young leaves emerge from the bud. With leaf expansion, these crystals move apart and continue to grow, while new crystals appear in the gaps between older, larger crystals. Later crystal patterns consist of evenly spaced, larger crystals at low density, or have clusters of small crystals in the gaps between large crystals attaining diameters of 40 to 50 ϕm in mature leaves.

Crystal spacing patterns appear to arise by the following mechanism: As apoplastic [Ca^{2+}] rises with transpiration, a few mesophyll cells (<0.5%) are induced to differentiate into crystal cells. As Ca^{2+}-sinks, crystal cells inhibit crystal cell induction in their vicinity by lowering apoplastic [Ca^{2+}] and thus prevent close spacing of crystals. During leaf expansion, new crystal cells are, therefore, induced only when gaps arise between the inhibition zones of existing crystal cells. Later changes in crystal patterns indicated increasing sink strength of crystal cells and temporarily increased Ca^{2+}-influx due to high transpiration rates.

Throughout leaf development, crystal cells apparently remain spaced at densities sufficient to maintain low concentrations of free apoplastic Ca^{2+}. The gradient between apoplastic and cytoplasmic $[Ca^{2+}]$ in the mesophyll and other plant tissues may thus be less steep than commonly assumed.

EFFECTS OF Ca^{2+} ON UBIQUITIN-INDUCED BREAKDOWN OF CALMODULIN

I. B. FERGUSON AND B. VEIERSKOV

DSIR Fruit and Trees, Mt. Albert Research Centre,
Auckland, New Zealand
Permanent address: *Department of Plant Physiology, Royal Veterinary and Agricultural University, Copenhagen, Denmark* (B.V.)

Ubiquitin is a small protein capable of regulating protein turnover. It is widespread in most animal tissues, and in lower organisms, both as a constitutive and induced protein. It is also present in plants, although the extent of its activity has not been widely investigated. Its mode of action is through covalent binding to target proteins, which are then either subject to proteolysis, or possibly to modification of existing enzyme activity.

In work on ubiquitin and regulation of protein breakdown during senescence of green plant tissue, we found that increasing concentrations of Ca^{2+} reduced the level of ubiquitin-protein conjugation. This was found both in whole tissue homogenates, and in chloroplast extracts, where we also have evidence for ubiquitin-protein conjugates.

In further work, specifically on calmodulin, increasing Ca^{2+} concentrations reduced the appearance of calmodulin breakdown fragments resulting from proteolytic activity induced by ubiquitin. Ubiquitin has been shown previously, in animal tissues, to bind to calmodulin, although the role of ubiquitin in regulation of calmodulin levels or activity has not been made clear. We have found that ubiquitin will reduce calmodulin-dependent Ca^{2+} uptake by microsomal vesicles. The effect of ubiquitin here is presumably due to calmodulin breakdown, or inactivation.

Ca^{2+} may be affecting ubiquitination of proteins either by inhibiting conjugation, or subsequent proteolysis. This is still unclear. However, it is likely that Ca^{2+} affects the structure of the ubiquitin/protein/protease complex, and its effectiveness will differ for individual target proteins. These results suggest a further role of Ca^{2+} in regulating cellular metabolism, particularly during senescence or in the plant's response to stress.

MATHEMATICAL BASIS OF CELL WALL SHAPE

S. BARTNICKI-GARCIA, F. HERGERT, AND G. GIERZ

Department of Plant Pathology and Department of Mathematics and Computer Science, University of California, Riverside, CA 92521 USA

Equation

The basic morphology of a fungal cell—that of a tip-growing hypha—can be described by the equation:

$$y = x \cot \frac{V \cdot x}{N}$$

where N = amount of wall-destined vesicles released from the VSC (vesicle supply center) per unit time; V = rate of linear displacement of the VSC.

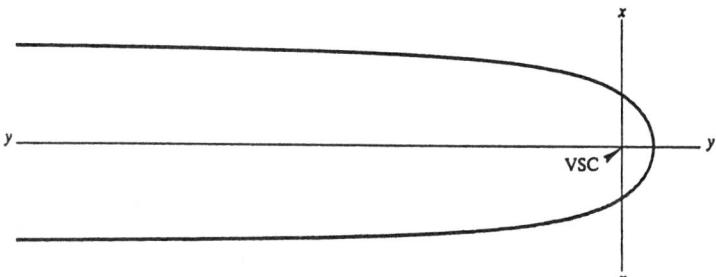

Assumptions

This equation was derived from a model of cell wall growth and morphogenesis based on vesicles. A computer simulation of the model was programmed with the following assumptions: (*i*) the wall is generated from materials discharged by secretory vesicles; (*ii*) vesicles are released from a postulated vesicle supply center; (*iii*) individual vesicles move from the VSC to the cell surface in any random direction; and (*iv*) the VSC is a mobile entity that can be displaced during cell growth (along the y axis).

Conclusions

(*i*) The position and movement of the VSC are critical determinants of morphogenesis. (*ii*) If the VSC remains *stationary*, vesicles reach the cell surface in about equal numbers in all directions, and the cell grows as a sphere. (*iii*) If the VSC *advances linearly*, while releasing vesicles, it generates the typical tubular shape of a hypha. (*iv*) There is a remarkable coincidence between the position of the VSC in the model and the position of the Spitzenkörper in real hyphae. (*v*) By manipulating V and N, a wide variety of fungal shapes can be simulated: spore germination, sporangium formation, yeast cell development, etc.

CHARACTERIZATION OF CALCIUM TRANSPORTING ENZYMES FROM MAIZE ROOT MEMBRANES

David Brauer, C. Schubert, and S-I Tu

ERRC/ARS, United States Department of Agriculture, Philadelphia, PA 19106, USA

The existence of a Ca-translocating ATPase in microsomes from maize roots was evaluated using assays to follow Ca-stimulation of ATP hydrolysis and Ca transport by changes in the fluorescence of chlorotetracycline (CTC). By following the distribution of Ca-stimulated ATPase activity by differential and density centrifugation in the absence and presence of free Mg, most of the Ca-ATPase activity was localized to the endoplasmic reticulum. Changes in CTC fluorescence were found to conform to a kinetic model which describes net transport as the difference between transport by the pump and inhibitory processes including feedback inhibition and ion leakage in the direction opposite of the pump. Calmodulin enhanced the initial rate and extent of change in CTC fluorescence. When these data were analyzed by the kinetic model, no change in the rate constant describing the processes that hinder the development of the ionic gradient could be attributed to calmodulin. Therefore, calmodulin affects the Ca pump directly. Since ATP hydrolysis was stimulated to about the same extent as changes in CTC fluorescence, calmodulin was probably increasing the turnover of the ATPase, rather than changing the stoichiometry between ATP hydrolysis and transport. When the Ca pump was inhibited by either the addition of vanadate, or the depletion of Ca by the addition of EGTA, the Ca gradient dissipated rapidly with the initial efflux conforming to first-order kinetics. In fact, the rate constant for the dissipation of the steady-state Ca gradient was greater than the rate constant describing the processes hindering the formation of the Ca gradient. Such results are possible only if active transport by the Ca pump reduced the membrane's permeability to Ca. The existence of Ca-proton transporters on plasma membrane and tonoplast has been investigated by examining the effects of Ca on the ability of the two pumps to form proton gradients as reflected by changes in the spectral properties of acridine orange. To date, we have found no evidence that proton gradients generated by the plasma membrane and tonoplast are linked to Ca transport.

DEMONSTRATION OF HIGH CALCIUM LEVELS IN SYNERGID CELLS OF ANGIOSPERM OVARIES

RAJENDRA CHAUBAL AND BONNIE J. REGER

ARS, United States Department of Agriculture, Richard B. Russell Agricultural Research Center, Athens, GA 30613, USA

In most angiosperm ovaries, the two synergid cells which flank the egg appear to regulate the following processes during fertilization: (*i*) chemotropic guidance of pollen tube growth toward the micropyle and into one (usually degenerating) synergid cell, (*ii*) arrest of pollen tube growth, and (*iii*) rupture of pollen tube tip to release sperm near the egg (4, 9).

Although identity of the putative chemotropic agent(s) and the cause of pollen tube arrest and rupture remain unknown, several studies on *in vitro* pollen tube growth suggest a role of calcium ions. Pollen tubes of some plant species exhibit directional growth toward increasing calcium in culture media (6). Supraoptimal calcium levels in these media can result in pollen tube arrest and rupture of their tips (2, 7, 8). Could calcium gradients in the vicinity of synergids be controlling sperm delivery processes of angiosperms? Only a few attempts have been made to demonstrate calcium gradients in angiosperm ovaries; none have specifically demonstrated the presence of calcium or measured its concentration in the synergids (3, 5, 1).

We fixed mature unpollinated wheat, pearl millet, rye, and maize ovaries by rapid-freezing in liquid propane (-189 to -191°C). Frozen ovaries were dried by substitution in cold acetone (-80°C for 48-72 h, -20°C for 4 h, 4°C for 1 h, and finally, room temperature for 30 min), critical point dried in liquid CO_2, and mounted on carbon stubs with adhesive carbon cement. Samples were viewed using a dissecting microscope, sectioned sagittally with a razor blade to expose the embryo sac, coated with carbon in a vacuum evaporator, and observed with a Philips 505 scanning electron microscope (SEM).

Elemental analysis of selected samples was performed using an energy-dispersive X-ray microanalysis system (Tracor Northern TN 5500 MCA) interfaced to the SEM. Electron induced X-rays were detected by Si(Li) spectrometer detector (Tracor Northern 3.0 mm^2 Microtrace) with a 7.5 μm-thick beryllium window, placed 20 mm from the specimen. The SEM was operated at 10 kV with stage tilt adjusted to obtain about 48.9 degrees take-off angle. Color-coded maps, depicting relative distribution and concentration of the elements, detected (Mg, P, S, Cl, K, and Ca) were generated using the Image Processing Program of Tracor Northern.

Within different regions of the ovaries, only the two synergid cells consistently displayed very high concentrations of calcium. Besides calcium, the synergids also contained some potassium and magnesium. The egg cell contained mainly potassium. Some calcium was also detected in the nucellus (wheat) and in the ovary wall (pearl millet).

Our results demonstrate that along the pollen-tube pathway in angiosperm ovaries, calcium is present in highest concentrations at the ultimate destination (synergids). Because synergids secrete substances into the micropyle (probably accompanied with initiation of degeneration of one synergid cell), formation of a localized calcium gradient in the vicinity of synergids is a realistic option. So far, however, we have not detected such a gradient in unpollinated ovaries, probably due to detection limits set by the instrument and uneven specimen topology. We are now using pollinated ovaries to investigate whether a (detectable) calcium gradient is formed after one synergid cell completely degenerates prior to pollen tube arrival at the micropyle.

LITERATURE CITED

1. Glenk HO, Wagner W, Schimmer O (1971) In J Heslop-Harrison, ed, Pollen Development and Physiology. Butterworths, London, pp 255-261
2. Herth W (1978) Protoplasma 96: 275-282
3. Jensen WA (1965) Am J Bot 52: 238-256
4. Jensen WA (1973) Bioscience 23: 21-27
5. Mascarenhas JP (1966) Protoplasma 62: 53-58
6. Mascarenhas JP, Machlis L (1962) Nature 196: 292-293
7. Reiss H-D, Herth W (1982) Planta 156: 218-225
8. Steer MW, Steer JM (1989) New Phytol 111: 323-358
9. Van Went JL, Willemse MTM (1984) In BM Johri, ed, Embryology of Angiosperms. Springer, New York, pp 273-317

PHOSPHORYLATION OF MEMBRANE POLYPEPTIDES OF BARLEY ROOTS

J. E. Garbarino and F. M. DuPont

*ARS, United States Department of Agriculture,
Western Regional Research Center, Albany, CA 94710, USA*

Phosphorylation of polypeptides in the membranes of barley roots was characterized *in vitro* and *in vivo*, and the effect of calcium on phosphorylation was investigated. Protein kinase activities were characterized *in vitro* in well-separated ER, tonoplast, and plasma membrane (PM) fractions. The isolated membrane fractions were incubated with $\tau\text{-}[^{32}P]ATP$, and the proteins separated by SDS-PAGE. *In vitro*, several proteins were observed in the tonoplast or ER fractions that were not also observed, at much greater intensity, in the PM fraction. Most of the phosphorylation was greatly enhanced in the presence of 90 μM calcium.

To measure phosphorylation *in vivo*, roots were treated with $[^{32}P]$orthophosphate and PM, and tonoplast fractions were isolated. Most of the phosphorylated polypeptides were present in the PM. Phosphoprotein bands appeared at 46 and 28 kD in both the *in vivo* and *in vitro* treatments. Many of the phosphorylated polypeptides, however, were different *in vivo* than *in vitro*. Phosphoproteins specific to the tonoplast fraction were observed at approximately 73, 66, and 50 kD. To test whether any of the phosphorylated polypeptides were subunits of the tonoplast or PM ATPases, blots of the phosphorylated proteins were probed with antibodies to the 72- and 58-kD subunits of the tonoplast ATPase, and to the 100-kD subunit of the PM ATPase. We also tested whether the calcium ionophore A23187 or salt shock could induce the phosphorylation *in vivo* of the polypeptides that were phosphorylated by a calcium-stimulated kinase *in vitro*.

CALCIUM IONS ARE INVOLVED IN THE ELICITATION OF LIGNIN BIOSYNTHESIS BY PECTIC FRAGMENTS IN CASTOR BEAN CELL CULTURES

Robert J. Bruce and Charles A. West

University of California, Los Angeles, CA 90024, USA

Pectic fragment elicitor (PFE) has recently been shown to elicit the rapid biosynthesis of lignin in suspension cultures of castor bean (1). This response is considered to be a mechanism of disease resistance which would help to limit the spread of a pathogenic organism in an infected plant. Concomitant with the onset of lignin deposition is the appearance of three new extracellular isoenzymes of peroxidase. These isoperoxidases may play specific roles in the biosynthesis of "disease-lignin." Calcium has been found to be essential for this elicitation to occur. Compared to cultures provided with sufficient calcium (2.5 mM), cultures grown in medium containing low levels of calcium (<1.0 mM) do not make lignin, or express any of the "disease-specific" isoperoxidases when treated with PFE. Additional research indicates that calcium is essential only during the elicitation interval. Pectic substances are known to bind calcium ions tightly. These results might imply that the elicitor-active molecule is actually a calcium-pectic fragment complex or chelate. Alternatively, PFE may stimulate the movement of extracellular calcium into the cell. This intracellular calcium could serve as a "second messenger" of infection to the plant. Studies under way to test these hypotheses using the calcium ionophore A23187 will be presented.

ACKNOWLEDGMENTS

Supported by USDA Grant 86 CRCR 12145.

LITERATURE CITED

1. Bruce RJ, West CA (1989) Plant Physiol 91: 889-897

STUDIES ON SIGNAL TRANSDUCTION IN PLANT DEFENSE GENE ACTIVATION

CHRISTIANE COLLING, KLAUS HAHLBROCK, HEINRICH KAUSS, AND DIERK SCHEEL

Max-Planck Institut für Züchtungsforschung, D 5000 Köln 30, Federal Republic Of Germany (C.C., K.H., D.S.) *and Universität Kaiserslautern, D6750 Kaiserslautern, Federal Republic Of Germany* (H.K.)

Cultured parsley cells (*Petroselinum crispum*) (3) and protoplasts (2) respond to UV light and a crude elicitor of *Phytophthora megasperma* f. sp. *glycinea* (Pmg) with the rapid transcriptional activation of phenylpropanoid pathways, resulting in vacuolar accumulation of UV-protective flavonoids and excretion of furanocoumarin phytoalexins, respectively. The mechanisms of signal perception and transduction leading to this specific gene activation are unknown.

In order to identify components involved in these processes, we used a monoclonal antibody against cyclic AMP (cAMP) in a radioimmunoassay (1) to determine the levels of this compound in parsley cells and protoplasts within the first 2 h of UV irradiation or elicitor application. The amounts of cAMP remained constant in protoplasts (3 ± 2 pmol/1×10^6 protoplasts) and cultured cells (6 ± 3 pmol/g fresh weight) after both treatments. Therefore, cAMP appears not to play a role as second messenger in these induction processes.

Omission of calcium from the culture medium of cells and protoplasts dramatically reduced elicitor-stimulated furanocoumarin excretion, whereas UV-induced flavonoid accumulation was not affected. These results were confirmed by nuclear runoff experiments. Elicitor-responsive genes were only fully activated in the presence of external Ca^{2+}, whereas UV induction of genes was not dependent on Ca^{2+}. Selective H^+-, K^+-, and Ca^{2+}-electrodes were used to measure the concentration of these ions in the medium. Within 2 to 3 min of fungal elicitor application, a decrease of Ca^{2+}- and an increase of K^+-concentration, as well as an alkalization of the medium, could be measured. Experiments with radioactive tracer showed a $^{45}Ca^{2+}$-influx within a few minutes of addition of *Pmg* elicitor. Thus, ion fluxes appear to be correlated with the elicitation process. Patch-clamp studies with parsley protoplasts and purified elicitor (4) will be performed to elucidate the involvement of Ca^{2+} and putative ion channels in the signal transduction pathway of the parsley defense response to *Pmg*.

LITERATURE CITED

1. Colling C, Gilles R, Cramer M, Nass N, Moka R, Jaenicke L (1988) Second Messengers and Phosphoproteins 12, 123-133
2. Dangl JL, Hauffe KD, Lipphardt S, Hahlbrock K, Scheel D (1987) EMBO J 6, 2551-2556
3. Kombrink E, Hahlbrock K (1986) Plant Physiol 81: 216-221
4. Scheel D, Colling C, Keller H, Parker J, Schulte W, Hahlbrock K (1989) *In* B Lutgenberg, ed, Molecular Signals in Microbe-Plant Symbiotic and Pathogenic Systems. Springer, Heidelberg (in press)

ENHANCEMENT OF MAIZE ROOT ELECTROTROPISM BY CALCIUM CHELATORS

H. Ishikawa and M. L. Evans

Department of Botany, The Ohio State University, Columbus, OH 43210, USA

We examined electrotropic curvature in primary roots of 3-d-old maize (*Zea mays* L., cv Merit) seedlings. The seedlings were mounted with roots immersed in oxygenated 1 mM MES buffer (pH 4.6-5.8, 24 ± 2°C) and arranged between two parallel, stainless-steel, plate electrodes across which a constant DC electric field could be maintained. The roots curved rapidly and strongly toward the anode (+). Both the latent period and rate of curvature were dependent upon the strength of the electric field. Electric fields around gravity-stimulated roots are known to be changed upon gravistimulation, and these changes may play a role in gravitropic curvature. We, therefore, compared some of the features of electrotropism and gravitropism in these roots. Gravitropism is known to depend upon an adequate supply of calcium, especially in the root cap. In addition, gravitropism is inhibited by auxin transport inhibitors and requires an intact root cap. We, therefore, examined the dependence of electrotropism on (*i*) calcium availability, (*ii*) auxin mobility, and (*iii*) the presence of the root cap. Electrotropism was enhanced by incorporating EGTA (0.1 mM) into the medium, or by decapping the roots. In contrast, treatment of the roots with auxin transport inhibitors (*e.g.* PBA), or with 10 μM IAA, inhibited electrotropism. Since gravitropism tends to counter electrotropism, we speculate that the enhancement of electrotropism by EGTA, or by decapping, results from the ability of these treatments to reduce or eliminate gravitropic sensitivity. The inhibition of electrotropism by auxin, or by auxin transport inhibitors, indicates that asymmetric growth in an electric field my be mediated by electrically induced auxin gradients.

CALCIUM TRANSPORT IN MEMBRANE VESICLES FROM BARLEY ROOTS

Frances M. DuPont, Douglas S. Bush, and Russell L. Jones

Department of Agriculture, Agricultural Research Service,
Albany, CA 94710, USA (F.M.D.) and
Department of Plant Biology, University of California,
Berkeley, CA 94720, USA (D.S.B., R.L.J.)

Uptake of $^{45}Ca^{2+}$ by membrane fractions from barley (*Hordeum vulgare* L. cv CM72) roots was characterized using membrane vesicles obtained from continuous and discontinuous sucrose gradients. Three Ca^{2+} uptake systems were detected, one being a Ca^{2+}-ATPase in the plasma membrane, one an unidentified Ca^{2+}-ATPase in the tonoplast-enriched region of the gradient, and the third being a Ca^{2+}/nH^+ antiporter in the tonoplast. However, no Ca^{2+} transport was specifically associated with the distinct peak of ER that was identified by NADH cytochrome c reductase, choline phosphotransferase, and dolichol-P-mannosyl synthase activities.

The main peak of $^{45}Ca^{2+}$ uptake was in the tonoplast region of the gradient, where the proton-translocating H^+-ATPase in the tonoplast vesicles supplied the energy for Ca^{2+} uptake via the Ca^{2+}/nH^+ antiporter. Sixty to 80% of Ca^{2+} uptake was dependent on the pH gradient. However, 20 to 40% of Ca^{2+} uptake was not dependent on the pH gradient and was inhibited by vanadate, indicating that a portion of the Ca^{2+} uptake was driven by a Ca^{2+}-ATPase. The unidentified Ca^{2+}-ATPase may be in the tonoplast, Golgi, or contaminating vesicles of unknown origin.

The coupled activity of the H^+-ATPase and the Ca^{2+}/nH^+ antiport in the tonoplast vesicles provided sufficient energy to drive the accumulation of 600- to 2,000-fold gradients of Ca^{2+} in the membrane vesicles. It is suggested that the same combination would be very effective at transferring Ca^{2+} from the cytoplasm to the vacuole *in vivo* with little effect on cytoplasmic or vacuolar pH.

The Na^+/H^+ antiporter is not observed in barley root tonoplast unless it has been activated by exposing the roots to salt. However, the Ca^{2+}/nH^+ antiporter was found in similar amounts in membranes from control and salt-grown roots.

ON THE ROLE OF CALCIUM IN THE ETHYLENE MEDIATED ABSCISSION OF CITRUS LEAVES

M. J. Jaffe and R. Goren

Biology Department, Wake Forest University, Winston-Salem, NC 27109, USA (M.J.J.); and Horticulture Department, The Hebrew University, Rehovot, Israel (R.G.)

Type II and type III explants were used to study abscission of 'Shamouti' and 'Clementina' citrus leaves grown in the field or the greenhouse, respectively. Calcium-channel modulators and EGTA were used as probes to study abscission, ethylene production, and calcium influx into abscission zones (AZs). The data is summarized in Table I.

Table I. *The effects of EGTA or calcium channel modulators on various properties of abscission zones.*

Modulator	Abscission	Ethylene evolution	Calcium Uptake
Gadolinium (10 μM)	accelerates	?	?
Nifedipine (50 μM)	accelerates	accelerates	inhibits
Verapamil (50 μM)	no effect	no effect	no effect
Bay-K8644 (50 μM)	inhibits	inhibits	?
EGTA (100 μM)	accelerates	?	?
EGTA + $CaCl_2$ (10 μM)	inhibits	?	?

? = not tested.

There are no compelling reports in the literature that directly demonstrate calcium channels in plant plasma membranes (PM). However, these data strongly implicate them. Calcium influx into the explants was measured as the decrease in murexide binding of calcium in the bathing solution. Using this method, calcium uptake is blocked by nifedipine, but not by verapamil. These results exactly parallel the effects of these calcium channel modulators on abscission.

We suggest that excision of the leaf induces opening of PM calcium channels in responsive cells. The resulting accumulation of free cytoplasmic calcium (FCC) tends to retard abscission of the explants. As the calcium channels close and FCC is pumped out of the cell, abscission is induced. Thus, gadolinium or nifedipine, which block animal PM calcium channels, accelerate abscission, whereas Bay-K8644, an agonist of animal PM calcium channels, retards abscission. Nifedipine increases and Bay-K8644 inhibits subsequent ethylene evolution. Because native ethylene action may induce cellulytic enzymes which participate in the second phase of abscission, FCC may modulate abscission via ethylene evolution.

INDEX

A-23187, 19, 47, 63, 98, 112, 114, 116, 117

A-9-C, 91

α-amylase, 60-65

Abscisic acid, 60-63, 148

Actin, 67, 68, 79-82, 151

Action potential, 69

Aequorin, 112, 116

Aleurone layers, 60, 62, 63

AM form of fura-2, 116, 117

Anaphase chromosome motion, 94-99, 101, 107

Antisense RNA, 128, 133

Apoplastic calcium, 5, 10, 12

Apple fruit, 1-6

Arsenazo-III, 95, 96, 98

ATP-dependent calcium transport, 154, 157

ATPases

    Ca^{2+}, 36, 46, 68, 72, 111, 132

    H^+, 36

    $Na^+ + K^+$, 36

    plasma membrane, 36

    vacuolar, 36, 55

ATPγS, 104

Autophosphorylating protein kinase (APK), 165, 166

Auxin (IAA), 168, 169, 174

BAPTA, 120, 121

BAY K-8644, 97

Bitter pit, 1-4

Black-heart, 3

Blossom-end rot, 3

Ca-related disorder, 3

Ca^{2+} binding, 30

Ca^{2+} displacement, 30, 33, 74, 75

Ca^{2+}-activated anion channels, 147, 148

Ca^{2+}/H^+ antiporter, 55

Caffeine, 69, 70

Calcimedins, 138

Calcium

    apoplast, 5, 10, 12

    binding proteins, 17-23, 137, 142

    bridges, 11, 12

    buffers, 49, 87, 89, 91, 97, 99, 101, 120

    cell wall, 11

    channel, 116, 120, 121, 159

    channel agonist, 97

    concentration, 17, 27, 93-98, 101, 103, 106, 107, 145-148, 168, 169

    current, 71, 75, 113

    gradients, 113-116, 121

ionophore, 97, 112, 114, 116, 151
membrane associated, 31, 94, 114
sensitive microelectrodes, 114, 153, 154
specific vibrating electrodes, 122
transport, 21, 48-50, 52, 61-63, 65, 154, 158
transients, 93, 96, 97, 108
Calmidazolium, 138
Calmodulin (CaM), 21, 38, 49, 50, 80-82, 127, 128, 130-134, 138, 139, 141, 142, 151, 162, 164, 166
binding protein, 127, 128, 130-134, 162-164
dependent kinase, 107, 164
sepharose affinity chromatography, 162, 163
Calsequestrin-like protein, 94
Caltractin, 140
Carbonylcyanide m-chlorophenylhydrazone, 47
Cardiac muscle, 67, 69, 75
Cardiac myocyte, 67, 69
Cation release channels, 148
Cell division, 93-97, 108, 137-142
Centrin, 138-142
Centrin-like protein, 138, 141, 142
Centrosomes, 138
Chara, 79, 80, 82, 86, 90
Chlamydomonas reinhardtii, 130, 132, 133

Chloroplast FBPase, 19
Chloroplast movement, 151
Chlorotetracycline (CTC), 94, 114, 116, 121
Chromosome motion, 95, 99, 102, 103, 104, 106, 107
Cork spot, 3
Cytokinesis, 97, 137, 140, 141, 142
Cytoplasmic streaming, 79, 81, 86-88, 90, 92, 98, 99
Cytoskeletal elements, 137
Cytosolic Ca^{2+}, 5, 7, 30, 32, 37, 95, 145-148, 168, 169
Donnan Free Space, 2, 14
Diacylglycerol (DAG), 76, 106, 107, 161
Dibromo-BAPTA, 101, 152, 153
Dihydropyridine (DHP), 75
Diltiazem, 97
E-C coupling, 74, 75, 91, 92
Elongation zone, 173, 174
Epidermal peels, 9
ER membranes, 38, 65, 93, 94, 121, 137, 140, 142
Erythrosin B, 50-52, 63
Exocytosis, 111, 115
Far-red light, 159
FCCP, 63
Ferredoxin, 20
Fertilization, 112
Fruit development, 2, 6

Fruit size, 2
Fucoid eggs, 111, 112, 120, 121
Fura-2, 114, 169, 173
G protein, 106, 111, 161
GDPβS, 104
Gibberellic acid (GA), 60-64
Gramicidin D, 47
Gravity, 86, 88-90, 92
Green alga, 151
GTP, 48, 50, 104, 106
GTPγS, 103, 106, 107
Guard cell, 141, 145
Heat shock (HS), 161-166
Heat shock proteins (HSPs), 161
Histone H_1, 162, 165
Hyperosmotic, 120, 122-124
Immunofluorescence, 81, 82, 138, 140
Indo 1-AM, 117
Indo-1, 98
Inositol 1,4,5-trisphosphate (IP_3), 5, 76, 104, 106, 156, 161
Inositol-(1,4)-bisphosphate, 5, 152
Ion channel, 130, 131, 144
K^+ uptake channels, 144-148
K^+/Na^+ selectivity, 26, 27
Kinetochore, 94, 108, 138
Lamins, 107
Lanthanum, 89, 97
Mass flow, 5

Membrane potential, 31, 69, 88, 91, 92, 148
Membrane vesicles, 62, 63, 65
Membranes, 29, 30, 33
Mesotaenium, 151, 154, 156, 158, 159
Metaphase, 94-99, 102
Metaphase, anaphase transition, 95-98, 107
Methoxy-verapamil, 97
Microfilaments, 161
Microtubules (MTs), 93-97, 103, 104, 107, 138, 140, 164
Mitochondria, 68, 69
Mitosis, 93-97, 108, 137-142
Mitosis-promoting factors (MPF), 107
Mitotic apparatus, 93-95, 98, 99, 103, 106, 107, 137
Mitotic regulation, 98, 107
Mitotic spindle, 139
Morphogenesis, 137
Mougeotia, 151, 158
Muscle contraction, 67
Mutant analysis, 128
Myofilaments, 67-70
Myosin, 67, 68, 80, 82
Myosin light chain kinase (MLCK), 131, 134
Na-pump, 69
Na/Ca exchange, 69-73
NAD kinase, 21, 132, 134
Neutral red, 89

Nifedipine, 89, 91, 97, 116, 117

Nitellopsis, 86-90

Normal Tyrode's solution, 72

Nuclear envelope breakdown, 96-99, 140

Osmotic gradients, 122

Paramecium, 128, 131, 133

Patch-clamp technique, 71, 144, 145, 149

Pelvetia, 112, 114, 122

Phleum, 174

Phosphatidylinositol 4,5 bisphosphate, 5, 106, 161

Phosphatidylinositol monophosphate, 5, 165

Phosphoinositides, 5, 32

Phosphoinositol (PI) pathway, 161

Phospholipase C, 76

Phosphorylation, 50, 80, 106, 107, 166

Phragmoplast, 95, 137, 139, 140

Phytochrome, 158

Plasma membrane, 32, 38, 47-50, 86-91, 161, 162, 165

Plasma membrane-associated protein kinases, 166

Plasmodesmata, 88

Polar axis formation, 112

Polarity, 140, 142

Polarized growth, 115

Pollen tube, 111, 115

Preprophase band, 140

Prophase/prometaphase transition, 96

Protein kinase C, 106, 107, 165

Protoplasts, 169, 173

Pyroantimonate, 94

Quin-2, 11, 97, 116

R 24571, 90

Red/far-red (R/FR), 151, 159

Relaxation currents, 146

Reticuloplasmins, 137

Rhizoid, 111-114, 120, 124

Root elongation, 168-174

Ruthenium red, 21, 22

Ryanodine, 69-74

Saccharomyces cerevisiae, 128

Salt stress, 26-33

Sarcolemma Ca-pump, 73

Sarcolemmal Ca channels, 67, 75, 76

Sarcomere, 67

Sarcoplasmic reticulum (SR), 67-76, 94

Schizoscharomyces pombe, 128

Secretory vesicles, 111, 121

Signaling agents, 104

Spindle pole, 94, 102, 139, 140

Spindle-associated membrane, 93, 94

Statoliths, 86

Stomatal movements, 144-148

Stretch-sensitive calcium channels, 122, 124

Telophase, 95

Tetraethylammonium Cl⁻ (TEA), 89

Thermotolerance, 161-164

Thioredoxins, 20

Tobacco cells, 162-164

Tradescantia, 96, 97

Troponin C, 67

Turgor pressure, 122, 124

Vanadate, 63

Vibrating calcium electrode, 120

Vicia faba, 144-148

Voltage-dependent anion conductance,
144-148

W-5, 90

W-7, 90

Wall loosening, 11, 12

Zea mays, 161, 168-170, 173, 174